IET CONTROL, ROBOTICS AND SENSORS SERIES 128

IoT Technologies in Smart Cities

IET International Book Series on Sensing–Call for Authors

The use of sensors has increased dramatically in all industries. They are fundamental in a wide range of applications from communication to monitoring, remote operation, process control, precision and safety, and robotics and automation. These developments have brought new challenges such as demands for robustness and reliability in networks, security in the communications interface and close management of energy consumption. This Book Series covers the research and applications of sensor technologies in the fields of ICTs, security, tracking, detection, monitoring, control and automation, robotics, machine learning, smart technologies, production and manufacturing, photonics, environment, energy and transport. Book Series Editorial Board

Proposals for coherently integrated international multi-authored edited or co-authored handbooks and research monographs will be considered for this Book Series. Each proposal will be reviewed by the IET Book Series Editorial Board members with additional external reviews from independent reviewers. Please email your book proposal to vmoliere@theiet.org or author_support@theiet.org.

IoT Technologies in Smart Cities

From sensors to big data, security
and trust

Edited by
Fadi Al-Turjman and Muhammad Imran

The Institution of Engineering and Technology

Published by The Institution of Engineering and Technology, London, United Kingdom

The Institution of Engineering and Technology is registered as a Charity in England & Wales (no. 211014) and Scotland (no. SC038698).

The Institution of Engineering and Technology
Michael Faraday House
Six Hills Way, Stevenage
Herts, SG1 2AY, United Kingdom

www.theiet.org

British Library Cataloguing in Publication Data
A catalogue record for this product is available from the British Library

ISBN 978-1-78561-869-7 (hardback)
ISBN 978-1-78561-870-3 (PDF)

Typeset in India by MPS Limited
Printed in the UK by CPI Group (UK) Ltd, Croydon

Contents

3   IoT-based smart water   **63**
Hitesh Mohapatra and Amiya Kumar Rath

4   Contiki-OS IoT data analytics   **83**
Muhammad Rafiq, Ghazala Rafiq, Hafiz Muhammad Raza ur Rehman,
Yousaf Bin Zikria, Sung Won Kim, and Gyu Sang Choi

About the editors

Fadi Al-Turjman is a professor and a research center director at Near East University in Nicosia, Cyprus. He is a leading authority in the areas of smart/cognitive, wireless and mobile networks' architectures, protocols, deployments and performance evaluation. He has served as the Lead Guest Editor in several journals including the *IET Wireless Sensor Systems* (*WSS*) and *Elsevier Computer Communications*. He was also the publication chair for the IEEE International Conf. on Local Computer Networks (LCN'18).

Muhammad Imran is an assistant professor at the College of Computer and Information Sciences, King Saud University (KSU), Saudi Arabia. He is also a Visiting Scientist with various universities including Iowa State University, USA. He is Editor in Chief of *EAI Transactions on Pervasive Health and Technology*, and serves as an associate editor for several journals including *IET Wireless Sensor Systems*, *IEEE Communications Magazine* and *Wireless Communication and Mobile Computing Journal*.

Foreword and outline

By Fadi Al-Turjman

The Internet of Things (IoT) is a system of connected computing devices, mechanical and digital machines, objects and/or people that are provided with unique identifiers and the ability to transfer data over a network without requiring human-to-human or human-to-computer interaction. One of the hottest areas in technology right now is, without a doubt, the smart city and sensors in the IoT era. Smart city and sensing platforms are considered one of the most significant topics in IoT. Sensors are at the heart of this concept, and their development is a key issue if such concept is to achieve its full potential. This book is dedicated for addressing the major challenges in realizing the smart city paradigm and sensing platforms in the era of cloud and IoT. Challenges vary from cost and energy efficiency to availability and service quality. To tackle such challenges, sensors have to meet certain expectations and requirements, such as size constraints, manufacturing costs and resistance to environmental factors existing at deployment locations. The aim of this book is thus to focus on both the design and implementation aspects in smart cities and sensing applications that are enabled and supported by IoT paradigms. Accordingly, our main contributions in this book can be summarized as follows:

- We start by overviewing the main enabling and emerging technologies in addition to security and privacy challenges for smart cities. Ensuring authorized accesses to these constituents is critical for ensuring the security in smart cities. We present existing solutions to these challenges mainly from the perspective of access control (AC) with a special focus on risk management, trust, insider threats and secure interoperation. We also present the associated future research directions.
- We discuss what is the role of IoT in smart cities describing the basics of what is IoT and what comprises a smart city followed by smart city segments. Benefits of IoT and their impact on the smart city along with the national and international case studies are also discussed. Challenges associated with the IoT with respect to smart cities are highlighted as well.
- We also focus on the role of the IoT in the smart water conversion process. Several problems like leakage detection, efficient water distribution and remote water monitoring can be addressed by using IoT with the combination of information and communication technology. In this line of thought, we propose a Smart Water Solution (SWS) architecture to ensure the proper utilization of natural or man-made resources. The hygiene water is the birthright of every human being, and to ensure it for a future generation, the SWS is a proven model.

- In order to help researchers, an evaluation tool which gathers information, analyses and develops simulation log results has been proposed. It provides detailed individual mote statistics as well as complete IoT network statistics. We compared various algorithms for Contiki-Operating System data analysis. Experiments show that all the proposed algorithms accurately calculate the overall metrics. However, adopted techniques for data analysis can make huge difference.

- As for the mesh IoT in smart cities, detailed analyses about the security aspects have been proposed. After entering the IoT mesh theme, a general review of IoT network security is presented. Next, security aspects in the IEEE 802.15.4 standard were analysed. Significant attention was also paid to the technical guidelines that enable secure transmission of information between selected IoT mesh points. The safety of implementing the mesh IoT network has been analysed in detail. Finally, the security aspects of different systems were compared.

- Moreover, IoT is used for detection of the location of the sediment deposition within the drainage pipe system to alert for repairing before complete blocking. However, from the hydraulic point of view, it is reasonable to design the drainage and sewer pipes to prevent the deposition of the sediment based on the physical parameters. To this end, instead of detection of blockage location, monitoring the flow characteristics is of more importance to keep pipe bottom clean from sediment deposition. Accordingly, solutions based on smart sensors mounted in the drainage and sewer pipes are discussed. Their ability to read the flow velocity and alert once the flow reaches a velocity in which sediment deposition is occurred has been also utilized in an intelligent method.

- Forthcoming smart cities can highly benefit from the existing online e-health systems to enhance Ambient Assisted Living. We present the opportunities and the challenges that come with this type of data and the applications while emphasizing the statistical analysis' importance on these topics. In addition, we show the steps of analysis based on a real data set obtained from binary sensors deployed in a smart home.

- Despite the significant advantages of the IoT medical device technology brings into the smart city healthcare, medical IoT devices are vulnerable to various types of cybersecurity threats, and thus, they pose a significant risk to the Smart City patient safety. Based on that and the fact that the security is a critical factor for the success of smart city healthcare services, novel security mechanisms against cyberattacks of today and tomorrow on IoT medical devices have been discussed as well.

- Moreover, the existing IoT devices registration mechanism in Wi-Fi for low power is based on carrier-sense multiple access with collision avoidance, which is not very efficient for the registration of large-scale IoT devices. Accordingly, we overview two potential authentication mechanisms, namely Centralized Authentication Control and Distributed Authentication Control, in detail. Later, we discuss another authentication mechanism, known as hybrid slotted-CSMA/CA (HSCT), as a case study that is proposed to overcome the aforementioned methods issues. The HSCT mechanism allows IoT systems in smart cities to register thousands of low-power IoT devices (sensors and actuators). This chapter also comes up with the analyses of the access period in a single HSCT time slot.

- Additionally, we apply statistical analysis on real data sets from 'Intel Berkeley Research Lab'. These data sets include 35 days of Mica2Dot sensor readings and voltage values. The main objective is to analyse the effects of environmental variables such as temperature and humidity on the lifetime of a sensor node. Common statistical models like regression and ordinal logistic regression have been used, and obtained results are discussed in detail.

Chapter 1

Access control approaches for smart cities

Nuray Baltaci Akhuseyinoglu[1] and James Joshi[1]

As the world population grows along with increase in urbanization, cities are getting more and more populated and increasing demands on various natural and man-made resources. Making cities smart through the appropriate application of the plethora of innovative new technologies and paradigms would help mollify potential environmental problems/constraints in such cities. Besides their significance in solving environmental problems, smart cities also aim to improve citizen's quality of life and efficiency of public services by optimizing the costs and resources involved. Smart city applications are founded on various enabling technologies and processes related to communication and networking, real-time control and big data analytics, to name a few. These constituents of a smart city infrastructure need to be integrated appropriately and seamlessly for provisioning efficient services to citizens. Each enabling technology has its own unique properties. The interplay and interaction among smart city constituents and their unique properties raise unique security and privacy challenges. In this chapter, we overview enabling and emerging technologies and security and privacy challenges for smart cities. Ensuring authorized accesses to these constituents is critical for ensuring the security of smart cities. We present existing solutions to these challenges mainly from the perspective of access control (AC) with a special focus on risk management, trust, insider threats and secure interoperation. Finally, we present future research directions.

1.1 Introduction

World population is increasing rapidly, and recently, there has been a significant growth in urban populations due to fast urbanization. Urban areas are estimated to be populated by 70 percent of the world population in 2050 [1]. Today, 88 percent of greenhouse gases are produced and 75 percent of the resources and energy of the world is consumed by cities. Continued growth in urban populations can make the scenario worse in the next few decades. Enabling smart cities through integration of emerging technologies is a highly promising approach to addressing environmental problems and resource constraints. For example, smart agriculture can help fight

[1]School of Computing and Information, University of Pittsburgh, Pittsburgh, PA, USA

against deadly diseases like salmonella poisoning by inspecting quality of raw and processed foods [2]. Similarly, smart water management applications can help detect in timely manner an ongoing diffusion of bacteria and contamination in the water by making use of sensors so that the concerned authorities can be promptly alerted [2].

One of the enabling technologies of smart cities is the Internet of Things (IoT), which comprises billions of connected devices. According to the forecasts by Cisco, the number of such devices is expected to grow to 28.5 billion in 2021, with 3.6 devices per person [3]. IoT is a future communication paradigm that is a dynamic network of devices interacting with each other, users and applications and that requires low response times [4]. Some IoT devices are resource-constrained in terms of power, computation and storage, such as the ones in wireless sensor networks (WSNs) and especially in wireless body area networks (WBANs) [5]. This growing network of interconnected "things" is used to collect and exchange data via ubiquitously deployed embedded sensors and to monitor, analyze and deliver valuable and actionable information [6]. Interaction between physical devices is enabled via several technologies such as sensor networks, cellular networks and wireless networks [4]. Interconnection of the devices to a gateway or the Internet is provided through wireless communication technologies such as Bluetooth, Bluetooth Low Energy (BLE), ZigBee, Wireless Fidelity (Wi-Fi), LTE-Advanced, Light Fidelity (Li-Fi) and 4G/5G [4,5]. The increase in the number of devices is expected to trigger the increase of demand in wireless communications because of their consistency in good quality and performance [4]. 5G is an emerging new generation mobile communication technology and is considered as one of the most reliable and efficient, as well as the fastest alternative for connecting billions of devices in IoT environments [6].

Smart cities have complex ecosystem and security and privacy requirements. However, IoT is far from meeting these security and privacy requirements because IoT devices are designed to focus more on meeting the goals of applications [7]. The growing number of IoT devices and their use in critical smart city applications, such as health-care systems and critical infrastructures, will have destructive impacts on the safety and security of the citizens and economies of the cities in case of a security breach. Similar to traditional information systems, smart cities are open to several security and privacy threats. Yet, the impact of security incidents in smart cities can be much more severe as the large, interconnected physical infrastructures and information and communication technology (ICT) systems are incorporated as key components in their design [2]. In smart cities, the result of security breaches may not only be broken or malfunctioning applications and services, but they can also significantly impair an entire city [2]. One example of a real-life experience is the attack that occurred in October 2016 which was targeted at one of the domain registration companies that controls a huge part of the domain name system (DNS) infrastructure [8]. It was a distributed denial of service attack mounted using a newly seen cyber weapon, Mirai botnet. The attack is presumed to be the largest of its kind that rendered a large portion of the Internet in the US inaccessible. What makes this attack devastating was that it utilized the ubiquitous IoT devices.

In this chapter, we first explore the smart city paradigm with various definitions and application areas. Next, we overview key enabling and emerging technologies of smart cities. We then discuss privacy and security challenges of smart cities and their unique characteristics that exacerbate these challenges. We review existing solutions for solving security issues of smart cities with a focus on AC. We then overview traditional AC models and frameworks to form a basis for our discussion about AC models and frameworks proposed for smart cities. We also review AC models that address presented security challenges in the chapter. Specifically, they are related to risk and trust management, insider attacks and secure interoperation issues. Finally, we discuss possible research directions for AC approaches for smart cities.

1.2 Smart cities

In this section, we present an overview of smart cities, smart city applications and key enabling technologies of smart cities. In Figure 1.1, we present a pictorial representation of smart city applications and enabling technologies.

1.2.1 What is a smart city?

There are various definitions of smart cities provided in the literature. Fernandez-Anez [9] surveyed definitions of smart cities from the aspect of stakeholders in a smart city infrastructure. They provide the following definition based on their analysis:

> A Smart City is a system that enhances human and social capital wisely using and interacting with natural and economic resources via technology-based solutions and innovation to address public issues and efficiently achieve

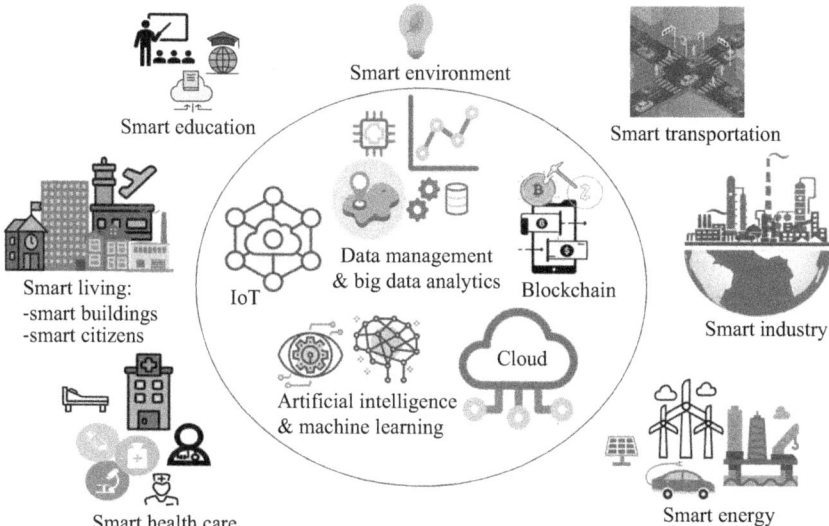

Figure 1.1   Smart city applications and key enabling technologies

sustainable development and a high quality of life on the basis of a multi-stakeholder, municipally based partnership.

Gharaibeh *et al.* [2] define smart cities from a data management perspective. According to them, a smart city comprises data management activities along with networking and computing technologies, which are protected using data security and privacy mechanisms. It encourages application innovation with the aim of elevating the quality of life for the citizens and has five dimensions: *utilities*, *health*, *transportation*, *entertainment* and *government*. As per the definition in [10], a smart city fosters collective intelligence of the city by fusing the social, physical, business and information technology infrastructures. Kondepudi *et al.* [11] define a smart city as

an innovative city that uses information and communication technologies (ICTs) and other means to improve quality of life, efficiency of urban operation and services, and competitiveness, while ensuring that it meets the needs of present and future generations with respect to economic, social, environmental as well as cultural aspects.

As can also be distilled from the definitions above, three main goals of smart cities are [2,9] *enhancing quality of life*, *achieving efficiency (or reducing cost of living)* and *obtaining sustainable environment*. According to Zanella *et al.* [12], the goals of smart cities are to utilize public resources better, to reduce operational costs of public administrators and to improve quality of the services provided to citizens. Stakeholders or end users of smart cities are defined as citizens, academia, intergovernmental institutions, local governments and private companies [2,9]. Each of them has different expectations from a smart city, in terms of capabilities and quality of service (QoS) [2].

1.2.2   Smart city applications

Smart cities consist of a collection of smart applications in several different domains. Gharaibeh *et al.* [2] discuss smart street lights, smart traffic management, virtual power plants in smart grid, smart health care and smart emergency systems as use cases of smart cities. In [13], smart city applications are grouped under five categories: *smart energy*, *smart environment*, *smart industry*, *smart living*, and *smart service*. According to Mohanty *et al.* [1], a smart city comprises smart infrastructure, smart transportation, smart energy, smart health care, smart education, smart citizens and smart technology. Zanella *et al.* [12] mention key industry and service sectors for smart cities as smart governance, smart mobility, smart utilities, smart buildings and smart environment. The most widely accepted categorization of smart city applications [9], which is also adopted by European Commission, maps smart city applications to one of the following six fields: *governance*, *economy*, *environment*, *mobility*, *people* and *living*. There are some cities which have realized services in these application domains, for example, Padova (Italy; proof-of-concept IoT island deployment), Stockholm (Sweden), Manchester (England), Hamburg (Germany), Barcelona (Spain), Amsterdam (Netherlands) and New York City (USA) [2,9,12,14].

Here, we briefly explain a subset of the smart city applications mentioned above. *Smart street lights* are used to cut down energy consumption in cities so that the saved energy is diverted to different areas like weather or pollution monitoring. This is achieved by dimming lights based on the time of the day or when there is low density of traffic or pedestrian activities [2]. *Smart grid* is a smart city application where smart meters are utilized for accurate measuring and billing purposes. There are aggregators in a smart grid that collect data from a group of houses in a residential area and send to control centers which in turn optimize power distribution [13]. Virtual power plants in a smart grid consist of a group of customers and are used by utility companies for purposes such as load reduction, demand response and pricing [2]. *Smart emergency systems* are used for purposes such as natural disaster management, dealing with accidents and crime detection and prevention in cities [2]. *Smart health-care applications* include a variety of components: in-body/on-body sensors, wearable devices, ICT, smart hospitals and smart emergency response to name a few. Smart hospitals utilize different technologies including cloud computing, advanced data analytics and mobile applications [2]. Patients' data are opened to access by multiple departments or hospitals to allow real-time sharing and decision-making on diagnosis and treatment of patients. *Smart traffic management systems* aim to efficiently monitor and manage the traffic flow of smart cities by reducing traffic congestion and response time to accidents and providing better travel experience to travelers [15].

1.2.3   Key enabling technologies for smart cities

Smart cities provide intelligent services by utilizing several different enabling technologies to achieve its main goals. These include IoT [1,2,12–14,16,17], cyber–physical systems (CPSs) [1,13,17], data management and analytics [2,15], blockchain [1,2,13], real-time control [13] and artificial intelligence (AI)/machine learning (ML) [2,18]. In this section, we discuss IoT, CPSs, data management, blockchain, cloud computing and AI/ML as key enabling technologies of smart cities.

1.2.3.1   IoT-enabled smart cities

IoT is a computing paradigm where devices with limited storage and computational capabilities are embedded in environments and communicate with each other via several wireless technologies [2,14]. Those constrained devices, like sensors and actuators, are known as "the Things." The Things are dispersed over large area [14] and massive in number [2]. They are used for the purpose of transmitting and receiving messages, collecting, storing and processing data ubiquitously, sensing and actuating [1,2]. On the other hand, the Things in the IoT may involve more sophisticated devices like smartphones, laptops, wearable devices and other entities like homes, vehicles, structures, buildings, energy systems, monument, landmark and appliances, which work collaboratively to provide services [1,14].

IoT plays a crucial role in the smart city infrastructure. IoT is a core component of smart city implementation and technical backbone of smart cities [1]. The goal of using IoT in smart cities is to enable more reliable applications with higher performance and improved security [14]. IoT has potential to furnish the intelligence, interconnection and instruments needed by smart cities [1]. Some IoT-related technologies for smart cities are radio-frequency identification (RFID) and WSNs, addressing of uniquely identifiable objects and middleware [14]. Some smart city services enabled by IoT include measuring structural health of buildings, waste management, air quality and noise monitoring, traffic congestion, city energy consumption, smart parking and smart lighting [12]. The main applications of IoT in smart cities are smart home, smart parking lots, weather and water systems, vehicular traffic, environmental pollution and surveillance systems [14].

From the perspective of the IoT technology, we can redefine the smart city infrastructure [13]. The smart city perceives physical changes from cities by incorporating ubiquitous sensors, heterogeneous network components and powerful computing systems, and acts back on the physical world. Ubiquitous sensing and real-time monitoring capabilities are offered by devices such as RFID devices, sensors and wearable devices. Data sensed from the physical environment are then sent to a control center using heterogeneous networks. Network infrastructure is considered to be heterogeneous because different networking technologies such as sensor networks, wireless local area networks (WLANs), wide area networks (WANs) and cellular networks can be integrated. A control center is supported by human operators and usually owns powerful computing resources used to process and analyze the collected data. Urban operations are manipulated based on the decisions made by the control center utilizing components such as actuators. Another description of the IoT [1] approaches it from the networking perspective, where the IoT is considered as a configurable dynamic global network of networks where the Things, local area networks, the Internet and the cloud are part of it.

1.2.3.2   CPSs, IoT and smart cities

CPS is another enabling technology of smart cities. In 2006, a working group of experts from the USA and the European Union proposed the term CPS [19]. The distinction between CPSs and the IoT is not apparent in the existing literature [1]. Some of them (e.g., [1,2,14]) consider IoT as comprised of physical structures (like buildings, bridges, landmarks and monuments), networking and communication infrastructure, and the Things as both constrained devices (like temperature sensor, microprocessor and microcontroller) and more sophisticated ones (like smartphones, wearable devices, microwave and washing machine). However, some others argue that the physical objects and sophisticated devices are not in the scope of the IoT [1]. This means that the IoT is a network of devices with limited capabilities whereas CPSs are built by using the IoT, physical components and sophisticated devices. As a result, CPS is supposed to be larger in scope compared to the IoT [1].

CPSs have a decentralized structure and the environment under which they operate is highly dynamic [20], much the same as the IoT. One distinctive feature of CPSs is that they are mostly safety-critical [21]. In addition, CPSs are usually emphasized to deeply embed cyber components into physical systems [22,23]. Even though embedded systems have existed for decades, CPSs have brought the focus on the integration of cyber components and physical systems [19]. This leads to an interaction between cyber and physical components, so called *cyber–physical interaction*, and *feedback loop mechanism* [19,22,24]. Enabling technologies for CPSs include multi-agent systems (MASs), service-oriented architecture (SOA) and cloud systems [19].

To illustrate the working dynamics of CPSs, here we use the medical CPSs (MCPSs) as an example. MCPSs bring together the medical devices with embedded software in them, networking facilities and abilities to capture complex physical processes of patients' bodies [25]. MCPSs consist of two types of devices, one of which measures the vital signs of patients (known as monitoring devices) whereas the second performs treatment-related tasks such as infusing the drug to the patient (known as delivery devices) [26]. In a typical scenario, delivery devices perform their scheduled tasks and monitoring devices monitor the patient's vital measures continuously. Usually a smart decision entity is included in the system which analyzes the collected data. For example, if the body temperature of the patient goes beyond the vital limits, then the decision entity generates an alarm to warn the physicians treating the patient. An example of such closed-loop system in patient care is a patient-controlled analgesia (PCA) system. In PCA, postoperative patients are delivered opioid drugs by PCA pumps for pain relief, overdose of which could cause respiratory depression and death. A monitoring device tracks the respiratory-related life-critical measures, namely heart rate and blood oxygen saturation levels, and the PCA pumps infuse the drug based on those readings. In case of a deviation from the normal readings, the PCA pump cuts the administration of PCA and prevents overdose or alert the clinician.

1.2.3.3 Data management in smart cities

Data management phases of a smart city are discussed in detail in [2,15]. The data management activities involve data acquisition, data processing and data dissemination [15]. Data acquisition is basically collecting data from resources. The technologies used for data acquisition are sensor networks, mobile ad hoc networks (MANETs), unmanned aerial vehicles, vehicular ad hoc networks (VANETs), IoT, social networks, 5G and device-to-device (D2D) communication [15]. Data processing corresponds to pattern discovery in the data. ML, deep learning and real-time analytics are related techniques. Data dissemination is the last step in data management process in smart city applications. Methods of data dissemination include direct access, push method, publish/subscribe services and opportunistic routing. In [2], Gharaibeh *et al.* point to the following four phases of data management: *(i)* data sensing and gathering, *(ii)* data fusion, processing and aggregation, *(iii) data exploitation* and *(iv)* service delivery. First phase corresponds to the data acquisition process discussed in [15]. Phases *(ii)* and *(iii)* together can be thought of as the data processing phase and Phase *(iv)* maps to the data dissemination phase mentioned earlier.

Another enabling technology under data management category is the data traffic analytics that is concerned usually with the analysis of data traffic behavior that can improve the quality of experience and QoS. An accurate data traffic modeling in smart cities and in the IoT environment is a requirement to prevent the energy waste [27,28]. Data traffic modeling corresponds to the provision of analytical models that are used to evaluate the performance of a communication systems, especially in terms of energy consumption [27,28]. Data traffic modeling approaches in the IoT environment can be classified into two categories as *static* and *dynamic* approaches [27]. Static approaches have an assumption that the number of mobile terminals and the base stations that they communicate with is fixed. Conversely, dynamic approaches capture the spatial and temporal fluctuations in the traffic load using system/network characteristics such as interference, capacity and spectral efficiency.

A data traffic model based on Markov discrete-time queuing system is proposed in [28] for wireless multimedia sensor networks (WMSNs), which are new generation sensor networks supporting real-time multimedia applications. Authors investigate the effects of multi-hop communication on Intelligent Transportation Systems using the proposed model. Simulations are performed based on realistic case studies to demonstrate the effects with respect to the following QoS factors: *average packet delay*, *energy consumption* and *network throughput*. As a result, optimal QoS parameters of forwarding are identified for multimedia routing protocols in WMSNs. Al-Turjman in [29] introduced information repeaters (IRs) as new components into an IoT network for reducing data traffic during the peak periods. IRs are sources of any redundant data such as in-network devices or extra devices placed close to the edge of the network in order to replicate the originally generated data. They formulate the problem of IR reallocation in high-demand regions as an Integer Linear Programming problem such that the total cost of content delivery is minimized while maximum network delay is limited. Also, they propose a new method to characterize the IoT network behavior (data requests or Internet traffic) based on the Content Demand Ellipses (CDE) model and analyze the effect of different traffic behaviors on the key parameters of CDE using real network traffic data.

1.2.3.4 Blockchain technology

Blockchain is a distributed public ledger formed by a peer-to-peer network of nodes. It has been widely known because of the popularity of Bitcoin [30], a cryptocurrency exchange platform [2]. Any transaction can be recorded to the public immutable ledger in a secure and anonymous way. To compromise the security of the blockchain, an attacker needs to acquire more than 51 percent of the total computing power of the Blockchain network [2]. Blockchain has been used in different applications including execution of financial transactions, cryptocurrency and smart contracts [31]. Smart contracts are self-executable programs such as in Ethereum [32] that can be used to custom code prespecified rules in the same way as in rules in a regular contract. Blockchain technology shows significant promise for smart city applications. A sample scenario for the use of blockchain in the smart city infrastructure could be applying smart contracts which mediate the relationship between service providers and citizens [2]. This would help to eliminate third parties and keep track of services or goods by users and payments by providers. In addition, blockchain can be utilized to solve some of the security and privacy challenges in smart cities [2].

1.2.3.5 Cloud computing

As explained in Section 1.2.3.1, control centers of smart city applications need to utilize powerful computational resources to process data and make informed decisions.

In this regard, cloud services provide significant benefits of storage and computational capabilities. However, when the cloud-based storage and computation are used, privacy-sensitive data of smart city applications are exposed to significant risks. The reason is that the cloud servers are not fully trusted and they are outside the control of users or organizations that use the them. Instead of keeping the data in a clear format and exposing to untrusted servers, encrypted data should be shared with or stored at cloud servers. On the other hand, this would introduce the challenge of processing encrypted data and analyzing smart city applications ineffectively and inefficiently [13]. To solve this, emerging encryption mechanisms such as homomorphic and functional encryption schemes are being developed [13]. There are two kinds of homomorphic encryption schemes: fully and partially homomorphic encryption. Fully homomorphic encryption schemes can be used by cloud servers for handling smart city application data, so that any kind of arithmetic computations can be performed on the encrypted data. However, they are not efficient and have high computational cost. Partially homomorphic encryption schemes could be preferred over fully encryption schemes because of better efficiency, but they support limited options for arithmetic computation [33].

1.2.3.6 Machine learning

Machine learning is a scientific study where ML algorithms are able to learn without being explicitly programmed [2]. In [2,5,18], ML, AI and big data analytics for smart cities and IoT are surveyed comprehensively. ML aims at developing self-taught computer programs that adapt to new data. These programs adjust their actions according to the patterns in the data that are inferred by ML algorithms. ML is categorized into three classes in general: *supervised, unsupervised* and *reinforcement learning*. A supervised learning algorithm uses labeled data, i.e. input data and corresponding outputs as labels, and infer patterns on a portion of the data known as training set. Later it estimates the output for newly added data by applying the patterns learned from the training data. In contrast, unsupervised learning algorithms do not rely on labeled data and discover the patterns hidden in the data. Decision trees, k-nearest neighbor, support vector machines and self-organizing map are examples of supervised learning algorithms whereas k-means as a type of clustering algorithm is an unsupervised method.

When simple patterns are to be analyzed, usually the use of basic classification and clustering algorithms is suitable. For analyzing more complicated patterns, *neural networks* are used since they outperform other ML methods [2]. At this point, the concept of *deep learning* comes to the scene. Deep learning is a type of ML and has become popular recently both in the academia and industry [2]. Deep learning comprises a set of ML algorithms that aim to model high-level abstractions from the data through linear and nonlinear transformations. At the core of deep learning lies neural networks that consist of a set of neurons and edges connecting them. Each neuron is associated with a "bias" value and each edge is labeled with a "weight." A neural network is structured in such a way that it consists of an input layer, an output layer and hidden layers in between. A disadvantage of neural networks is their scalability problem when the number of neurons grow. There exist various deep learning models including restricted Boltzmann machines and long short-term memory.

ML techniques can be used for different purposes in a smart city. They can be used either for network performance optimization/system optimization or for extracting meaningful information from the sensed data [5]. In system optimization category, they have been used to improve power consumption, routing of data and to predict sensor failures [5]. As an example to the use of ML for deriving meaningful information from the sensed data, they have been used in smart grids to transform historical electrical data into models and predict the risk of failure for components so that maintenance works can be prioritized [2]. They have been also used for traffic flow optimization, in water management systems, to estimate the energy efficiency of buildings, for developing safety applications and detecting traffic accidents in VANETs [2]. Deep learning algorithms have been used together with geo-spatial urban data to address different needs in smart cities. For example, there have been works that use deep belief network or recurrent neural network (RNN) for dividing metropolitan areas into zones based on travel patterns of residents, for smart water management networks, electricity load forecasting, real-time optimization of pricing in smart grids, traffic flow prediction and crowd density prediction in metro areas [2].

1.3   Security and privacy challenges

Since the information collected, stored and communicated in smart cities is sensitive, it is crucial to take careful measures against unauthorized access, disclosure, modification and disruption of information. This requires consideration of security and privacy requirements for different layers/components/applications of smart cities, such as confidentiality, integrity, availability, non-repudiation and privacy [13]. AC and authorization mechanisms provide a key foundation for addressing such security and privacy requirements. Authentication and secure communication protocols help address some security requirements in IoT environments by verifying the identity of users, such as the context-sensitive seamless identity provisioning framework proposed for Industrial IoT in [34]. In this section, we present a high-level overview of security and privacy challenges in smart cities. We also discuss the unique characteristics of smart cities that exacerbate these challenges.

1.3.1   Unique characteristics of smart city applications

Smart city applications have unique characteristics that introduce nontrivial security and privacy issues and render traditional solutions not very effective. They require designing security solutions that are specifically tailored to the needs of smart city applications. These unique characteristics of smart cities are:

1. integration of cyber and physical environments,
2. resource-constrained IoT devices, and
3. heterogeneity of devices and software components.

The first characteristic is that the physical infrastructure and ICT are intricately interwoven. This causes the impact of security incidents to be significantly magnified compared to that in other ICT systems [2]. Another unique characteristic that applies to IoT-enabled smart cities and arises from the nature of constrained devices is the limited storage and processing capabilities. This limits the use of state-of-the-art security and privacy mechanisms. For example, the state-of-the-art cryptographic algorithms cannot be implemented on resource-constrained IoT devices and this leads to choosing weaker options. Third dimension is the heterogeneity of devices [10] and technologies used in smart cities, which may cause incompatibility issues when different technologies and devices are integrated [2]. For instance, smart grids face the issue arising from the use of legacy systems and protocols. Legacy protocols, like the ones for Supervisory Control and Data Acquisition (SCADA) system in the smart grid, have been designed decades ago for efficiency purposes without considering security issues. In addition, legacy devices in smart grids do not support current security solutions like the ones based on cryptography and cannot be patched [2].

1.3.2 Privacy

The reason for the information in smart cities to be sensitive is that it is collected from peoples' environments and reflects their lifestyle, which in turn is processed and used to enable applications to support the population [13]. For instance, health conditions of a patient in a smart hospital, identity and location of a user in a smart transportation application, lifestyle of residents in a smart home or smart energy applications can be leaked in case of a privacy breach [13]. For example, let us think of an aggregator device in a smart grid [13]. This type of device aggregates metering data from several houses in a residential area. It may jeopardize the privacy of residents if it acts dishonestly by revealing the data it observed. Metering data are sensitive because it can reveal lifestyles of people through the energy consumption patterns over a period of time. According to the authors in [35], regardless of the type of a meter, whether electric or gas or thermal, personal data collected via meters in smart cities belong to end users and hence are subject to privacy control. As another example, we can consider smart transportation applications [13]. Smart transportation applications include a navigation service that could be provided dynamically to give information about the current state of the traffic, delays along a path to a destination, etc. This type of service requires real-time location information of the users' vehicles for querying dynamic road conditions, which poses a threat to privacy of users, in particular, to their whereabouts.

Zhang *et al.* in [13] propose a smart city architecture that consists of sensing components, heterogeneous networks, processing unit, control and operating components as explained in Section 1.2.3.1. Some of the available solutions, like encryption and anonymity mechanisms are recommended for protecting the privacy of the data handled during sensing. For protecting the privacy of network part of the IoT architecture, following privacy enhancing techniques can be deployed [36]: VPNs (virtual private networks), TLS (transport layer security), DNSSEC (DNS security extensions) for providing authenticity and integrity of the data origin, onion routing for hiding the source of an IP (Internet Protocol) packet, and PIR (private information retrieval).

1.3.3 Secure interoperation

A daunting security challenge for smart cities is that of secure interoperation among heterogeneous entities with different resource constraints and security requirements. Securing interoperation among these entities are critical to facilitate seamless information sharing or collaboration among interacting entities. As an example, let us consider smart traffic management or transportation systems. In those systems, the data about road traffic are collected using crowd sourcing approach via mobile devices of people traveling or via cameras deployed on roads. The challenge emerges during global road planning when AC policy is to be defined for enabling data sharing among collaborators in a privacy-preserving manner [13]. The need for collaboration among multiple parties is also critical when we approach smart cities from the perspective of the IoT and CPSs because they are typically a part of large-scale distributed systems, involving multiple subsystems, teams, departments or organizations. In addition, there is an emerging trend towards designing CPSs using SOA paradigm as proposed in [19,22,23]. This requires collaboration among and integration of different services through secure service orientation and composition to present the users with high QoS and to decrease the time and effort for developing services out of the expertise area of an organization.

In smart cities, multiple stakeholders come together and form collaborations to provide services to citizens. Each party would be represented as an individual domain with its own set of information resources and services, as well as security and privacy requirements. These heterogeneous domains need to interoperate to fulfill their goals of service provisioning through sharing of sensitive information. If we approach the smart city from the IoT perspective, the Things may belong to different domains, such as smart sensors of WSN domain, smart traffic signals of smart traffic management domain and smart home alarms of smart home domain [16]. An example of interoperation in smart cities could be that of the air pollution monitoring in smart cities, presented as a motivating scenario by Cao *et al.* [16]. In this sample scenario, the stakeholders involved in the collaboration and representing individual domains are the company which collects air pollution data via sensors, municipal authorities and commercial operators. The biggest concern making collaborating organizations hesitate to share their information and computing resources is ensuring security and privacy during their interoperation. Security is an absolute requirement in such interoperation environments due to the high possibility of existence of malevolent parties and the lack of trust among participating entities. This fact points to the problem of how to realize interoperation in a multi-domain environment in a secure and privacy-preserving way. The key to enabling secure and privacy-preserving interoperation in a multi-domain system is enabling privacy-aware inter-domain accesses through integrated AC policies and relevant mechanisms [37].

1.3.4 Insider attacks

Security and privacy mechanisms on hand are mostly envisaged for guarding against outsider attacks. However, smart city applications are also open to attacks which can be executed by potential adversarial insiders like employers and contractors, or a component within the supply-chain or a multi-domain environment [13]. They can put the organization or the smart city applications at high risk by stealing/exfiltrating users' privacy-sensitive information such as health-care records and blueprints of a smart city. Insider attacks have higher chance of being successful compared to external attacks as the adversary can be expected to have significant insider information of the smart city environment [38]. One reason for this is that insiders hold the necessary credentials to have access to devices and information resources inside the network of an organization. In addition, insiders are physically present in the proximity of the organizational devices. This allows more options for the actions to be performed by insiders, compared to outsiders. Another reason is that since insiders perform actions in the trusted network of the organization, their actions do not go through extensive security checks as an external access does. Furthermore, it may be easier for insiders to circumvent security controls as they have deeper knowledge and training about organizational network configurations, policies and implemented/deployed security mechanisms.

Insider attacks can be launched by authorized employees of organizations by abusing the trust intentionally for noxious purposes or caused unintentionally as a result of their imprudent behaviors. According to the findings of 2017 US Cybercrime report [39], which presents the results of a survey about security incidents that occurred in 2016 and self-reported by more than 500 organizations, insider attacks constituted 20 percent of the attacks experienced by organizations. In fact, insider attacks may result in higher costs compared to the attacks carried out by outsiders, as confirmed by 43 percent of the participating organizations in the survey. The effects of these attacks are manifold, varying from compromised records of customers (40 percent of the total incidents) and employees (38 percent) and confidential records such as trade secrets and intellectual property (33 percent). These in turn may cause severe damage to the reputation, revenue and customer loss. The key challenge for an organization when dealing with insider threats is to authorize users with the right set of privileges since insufficient privileges refrain employees from performing their tasks, whereas excessive privileges may be abused by malevolent users or misused accidentally [40]. Security mechanisms put into practice in an organization for external attacks may not be adequate for thwarting insider attacks. For example, leakage of data outside of the organization by insiders cannot be detected by anomaly detection systems or fire walls, because data are usually leaked through legitimate channels [38].

1.3.5 Trust management

The significance of trust is well appreciated in the scientific literature, yet there is no one definition of trust that is broadly agreed upon [41]. In smart cities, trust is usually approached from the IoT perspective as is done in [16,42]. Trust is an important issue to consider for the IoT environment because of the fact that the data collected from sensors are transmitted through the network, analyzed in real time and used for actuation. If trustworthiness of the data at any of these phases is infringed upon, then this signals a possible compromise and can prevent a proper functioning of the IoT devices that are part of a smart city. IoT has unique features that raise challenges with regards to the assurance of trust. These features include vulnerable transmission medium, untrustworthiness in IoT sensor data and ever-changing and heterogeneous network topology/devices [42]. While the first feature is self-explanatory, untrustworthiness in IoT sensor data arises from redundant data copies kept for availability purposes, and from faulty and compromised sensors. If an attacker is able to compromise IoT devices/sensors, then the attacker will be able to generate or report fake data. This will result in a disruption in the whole data management process of the IoT. Ever-changing network topology is a challenge which emerges as a result of the constant movement of IoT nodes.

There are several attacks that can jeopardize trust management in an IoT environment. Some examples are on-and-off attack, bad-mouth attack and ballot-stuffing attack [42]. In an on-and-off attack, the malicious behaviors are displayed in some period of time and then stopped so that finding the attack patterns by trust management schemes becomes more difficult. In a bad-mouth attack, an attacker targets benign devices and downgrades their trustworthiness by providing bad recommendations to hide the malicious behaviors of the compromised nodes. In a ballot-stuffing attack, an attacker works towards raising the trust of another malicious node via good recommendations to help it stay in the system longer. To tackle trust-related attacks in smart cities, different approaches can be followed. One approach is to evaluate the trust of IoT sensors and sensor-enabled devices and detect malicious ones, as proposed in [42,43]. Li *et al.* in [42] propose RealAlert, which is a policy-based and trustworthy sensing scheme for the IoT. In this scheme, the trust of both the data and the IoT devices are evaluated using the history of data reporting and the context under which the data are collected. More specifically, in case an IoT node starts to report abnormal data, first the contextual information is used to decide whether the reason is contextual factors or a deliberate dissemination by adversaries. In this way, malicious nodes are differentiated from the nodes that are benign but provide faulty readings. Some of the proposed work [44,45] handle an IoT environment as a layered architecture and build trust solutions embracing different layers, such as sensor layer, core layer and architecture layer. Other approaches adopt reputation-based approaches, fuzzy techniques, social networking-based approach and historical behavior-based approaches to trust management [41]. Last but not least, trust needs to be appropriately leveraged to ensure AC and secure interoperation in a smart city environment.

1.4 AC approaches for smart cities

In this section, we present AC approaches for smart cities. First, we briefly explain traditional AC models to build discussion about current approaches which extend them. Then, we present AC models and frameworks that have been proposed specifically for smart cities or smart city applications and their enabling technologies. Finally, we present AC approaches which target to alleviate the security challenges we discuss in Section 1.3.

1.4.1 AC models

Various AC models have been proposed since late 1960s to address security requirements of systems and applications. Among those, discretionary AC (DAC), mandatory

AC (MAC) and role-based AC (RBAC) models are the widely known ones. A good overview of these models has been presented by Samarati and De Capitani di Vimercati in [46]. DAC works by enforcing authorization rules that specify the actions allowed or forbidden on resources of a system and are written based on the identities of users. The reason to name the model as "discretionary" is that users can use their discretion to transfer the privileges they have to other users, whereas the authorizations are managed by administrative policies. Access matrix model, which is first proposed by Lampson [47] with the purpose of protecting resources in operating system contexts, is a means used to realize DAC. Lampson's model was refined later by Graham and Denning [48] and formalized by Harrison, Ruzzo and Ullmann (HRU model) [49]. Access matrix model consists of the following components: a set of subjects (S), a set of objects (O) and an access matrix (A), where the authorization state of a system (Q) is represented by a tuple (S, O, A). Subjects are the entities that are entitled with access privileges over the objects. Rows and columns of A denote the subjects and objects, respectively, whereas each cell ($A[s, o]$) includes privileges or rights. An authorization state of a system, Q, can be changed by issuing commands that consist of conditions and operations. HRU model defines six operations or primitive commands, which are add/delete subjects, add/delete objects and add/delete privilege (to/from $A[s, o]$). Because of the fact that A is usually sparse, it is implemented as authorization tables, AC lists, or capabilities in practice, for efficiency.

DAC policies are prone to vulnerabilities, specifically to Trojan Horses, which may lead an AC system to be bypassed by malevolent parties. This happens due to the fact that they do not consider distinction between users and subjects, where users are real identities trusted for their actions and subjects are processes, which execute operations on behalf of users relying on their authorizations. This causes malicious programs to be executed unbeknown to the user. Defending against this vulnerability requires enforcing controls on information flow enabled by processes, where MAC comes in. A MAC system enforces AC based on a central authority in contrast to DAC. One of the well-known approaches to MAC is multilevel security policies. The MAC approach handles authorizations by assigning subjects and objects to security *clearances* and *classifications*, respectively. Access classes consist of a *security label* and a *category* and are compared using a dominance relationship ($\geq$). *Security labels* reflect the sensitivity of the information contained in objects and the trust attributed to subjects, also known as *the clearance of subjects*. They have hierarchical order and can be defined as Top Secret > Secret > Confidential > Unclassified. *Categories* depend on the application domain, which can be Administration, Research and Financial for a commercial system. Security levels and the dominance relationship between them generate a security lattice in a multilevel security policy system. A security lattice is utilized to limit the access privileges of subjects to the ones required by their tasks, the concept which is known as *need-to-know*. Multilevel security policies can be defined based on two purposes: protecting confidentiality or integrity. The Bell LaPadula [50] model is an example to a confidentiality-based multilevel security model, where the information flow is controlled by using "no read-up and no write-down principles," so that the sensitive information is not leaked to unauthorized users via processes executed on behalf of the trusted users. Biba's model [51] is based on integrity, where the integrity of the information is protected by satisfying "no read-down and no write-up principles," so that the information only flows from higher or more reliable/trusted levels (or sources) to lower levels.

Because of the shortcomings of DAC and MAC models, specifically related to the administration burden, RBAC has emerged as a promising approach to address the needs of practitioners [52]. There are several proposals for RBAC in the literature and they are all established on the same principles. The RBAC model comprises the following components: *users, roles, permissions* and *sessions*. The concept of *role* is essential in an RBAC approach and it is associated with responsibilities specific to organizational duties. The scope of a role can range from the job title of a user, like purchasing manager, to a specific task to be fulfilled by the user, like purchasing order management. Access privileges are defined based on roles, instead of writing rules separately for each user. Users can obtain access privileges, i.e. permissions, authorized to the roles to which they are assigned by activating them in a session. A user can activate different roles assigned to him and a role can be activated by multiple users. A user can activate multiple roles simultaneously, is restricted to just one role or can activate multiple roles but with constraints on the activation of specific roles. A permission is actually a pair, *(object, access mode)*, where objects are protected resources against unauthorized access and access mode can be read, write, execute, etc. Authorization management in RBAC requires two types of assignment: assignment of users to roles and assignment of permissions to roles. This provides a significant reduction in the management burden on the administrators.

Role hierarchy, least privilege, separation of duties (SoD) and constraint enforcement are the features supported by the RBAC. A role hierarchy is based on *generalization* and *specialization* principle. A role hierarchy further eases the administration burden, due to *inheritance* from specific (junior) to general (senior) roles. In this way, a senior role acquires permissions that belong to its junior roles. *Least privilege* enforces limit on the authorizations provided to a user by means of a set of roles so that excessive privileges are not exercised. In [53], least privilege is provided via the use of role activation hierarchies, so that first, the junior roles are activated, and their permissions are acquired by users without activating senior roles and exercising excessive privileges. *SoD* principle is for preventing misuse of permissions by a single user. A well-known example to the SoD is separating the privileges of preparing a check and authorizing it. SoD is categorized as static and dynamic, where static separation (SSoD) requires not assigning conflicting roles to the same user and dynamic separation (DSoD) requires the control over the activation of conflicting roles by the user within a session. *Constraint enforcement* corresponds to the feature that lets an RBAC administrator specify constraints on role activations. For example, a constraint can be defined on the number of users that can activate a role simultaneously or the number of activated roles that are associated with a given permission, etc. A drawback of the RBAC is that the ownership relationship between a user and an object is not represented as in the DAC, because creating an object with a given role does not necessarily mean that the user owns the object.

RBAC saves the security administrators from the administration burden of access rights. Hence, the RBAC has been widely accepted and become a dominant AC model for the last 20 years or so [54]. Although the RBAC provides higher scalability than the DAC and MAC models for a larger number of users in an AC system by simplifying the management of access rights [26], several researchers have pointed out its limitations [52,54]. Researchers have proposed a huge number of extended versions of the RBAC to overcome various limitations [26,52]. These extensions have not been able to solve certain problems like the lack of scalability and responsiveness to the dynamic nature of real-time distributed systems [26]. Hence, there has been a growing interest recently in attribute-based AC (ABAC) [52,54]. ABAC is a more general and flexible AC model than the previous ones that can address many of the AC principles such as least privilege, SoD, etc. Therefore, any other AC model can be thought of as a special case of an ABAC model [37]. In other words, ABAC can be configured to function as MAC, DAC or RBAC as shown by Jin *et al.* [52]. ABAC is still immature and in its early development phase. NIST has published a guidance for the definitions and implementation of ABAC [55]. However, there is still no formal standard which is agreed upon and recognized by all. Here, we will explain the core ABAC model [54] that is formalized by Jin *et al.* [52].

ABAC uses the attributes of requesters and objects to grant access to objects. Owing to this working principle of ABAC, the identities of the users do not have to be known beforehand to the party sharing its resources with external parties. There are three finite sets of elements in the core ABAC, which are users U, objects O and actions A. Users, who can be humans or applications, ask for permissions to carry out actions on objects, which are resources to be protected from unauthorized accesses, where the set of actions includes traditional object-oriented operations, such as read, write, update, create, delete etc. $Perm(A, O)$ represents the permission provided to the user if certain conditions are met, based on the policies defined. Policies are written with respect to the finite set of attribute functions *Attr* that consists of a finite set of user attribute functions *UA* and object attribute functions *OA*. *AttrType* is a function that returns the type of an *attr* $\in$ *Attr* as "set-valued" or "atomic-valued." An attribute function, *attr* $\in$ *Attr*, maps a user or an object to a value (or multiple values in case of a set-valued attribute type) from the set of possible values, *Range(attr)*. *Auth* represents the authorization function that evaluates an access request based on the attribute values of the requesting user and returns a decision as permit or deny.

1.4.2   *AC models and frameworks for smart city and smart city applications*

In this section, we overview AC models and frameworks that have been proposed in the literature by considering applications or enabling technologies of smart cities. SMAR-TIE platform is proposed by Beltran *et al.* [7] for smart city applications to provide an efficient, decentralized, and user-centric authentication and authorization for IoT applications. This platform allows enforcing privacy preferences of users via policy-based AC and attribute-based encryption techniques. To design the architecture of the SMARTIE, authors take the IoT Architectural Reference Model (IoT-ARM) [56] as the baseline. Authors review different communication models for IoT applications, which include device-to-cloud, device-to-gateway and D2D models. These models differ in the AC architectures that they use. First two of them are centralized approaches. In the device-to-cloud model, AC is only to be handled between an IoT device and a service provider. In the device-to-gateway model, a gateway plays an intermediary role between an IoT device and the Internet. The gateway is responsible for AC and there is an AC list attached to it and updated by the cloud server. In the D2D model, IoT devices communicate with each other without any intermediary. This requires devices to manage authorizations themselves. The authors claim that centralized models, first two of the models, are infeasible as they require configuration and updates to the identity management and authorization policies of every device, which introduces scalability challenges. Instead, a decentralized D2D communication model with a decentralized AC mechanism is more applicable. To meet these requirements, i.e. providing user-centric privacy and AC, decentralized architecture and scalability, authors include an eXtensive Access Control Markup Language (XACML), capability-based AC (DCapBAC [57]) and Ciphertext-policy attribute-based encryption (CPABE) as AC-related components in the SMARTIE. XACML allows users to specify fine-grained authorization and privacy policies. DCapBAC server uses XACML authorization policies to generate lightweight JSON (JavaScript object notation)-encoded authorization tokens as proofs to be used by client applications when they request resources from the server. Here, the decentralization is applied via the use of sensors or actuators as authorization servers by deploying DCapBAC components in them. Data brokers are used for saving the resources of limited power devices like sensors. Clients communicate to data brokers to request sensor data. When a data broker is to communicate the sensor data with a client, it first requests the encryption of the data by CPABE component. Based on the XACML formatted privacy rules, CPABE encrypts the data and sends it back to the broker. Without further control on the data, the broker shares the encrypted data with subscribing clients.

Alshehri and Sandhu in [58] propose an AC-oriented architecture (ACO) for IoT environments with a purpose of developing a family of AC models for the cloud-enabled IoT. ACO is proposed based on the authorization requirements of the IoT and consists of four layers: object layer, virtual object layer, cloud layer and application layer ordered from lower to upper levels. Object layer comprises physical IoT devices and virtual object layer includes digital representations of the devices used for maintaining status of physical devices. Physical devices communicate with each other using communication protocols such as Bluetooth, Wifi, Zigbee and LAN. They communicate to virtual devices in the virtual object layer via communication protocols such as HTTP, MQTT, DDS and CoAP [59]. The need of AC is highlighted for upper layers, which are cloud and application layers. Bhatt *et al.* [60] propose a formal AC model for the IoT: Amazon Web Service IoT Access Control (AWS-IoTAC) model. AWS-IoTAC extends previously proposed Amazon Web Service Access Control (AWSAC) model that is a model proposed for cloud services in general. AWS-IoTAC incorporates AWS-IoT service and IoT related components into AWSAC, such as IoT devices (things), certificates (X.509 certificates), IoT operations, rules and AWS services. Authors map AWS-IoTAC model to the previously proposed ACO architecture [58]. They present two use cases to show how their model can be applied on the AWS IoT platform, utilizing lightweight MQTT protocol that provides connection and communication among IoT devices. They propose to attach

AC policies to X.509 certificates, then to attach those certificates to virtual objects and to copy the certificates to physical devices.

Gupta and Sandhu in [59] propose an AC framework for the Internet of Vehicles (IoV) and cloud-assisted connected cars. Cloud-assisted connected cars paradigm corresponds to the scenario where the connected cars in the IoV are supported by cloud systems. The proposed framework, so called the extended ACO (E-ACO), extends the previously proposed ACO architecture [58]. Similar to ACO, E-ACO consists of four layers: object layer, virtual object layer, cloud services layer and application layer. Virtual object is an essential element of the IoV and the E-ACO. The reason is that the devices in IoV are heterogeneous with respect to communication protocols they use and have high mobility, which cause connectivity problems. With the use of virtual objects, i.e. digitally maintained states of physical devices, the communication within and among cars can be provided even when the physical devices are not connected. The authors propose to keep virtual objects close to physical objects in E-ACO because high latency and low bandwidth problems are encountered in virtual object creation. E-ACO differs from the ACO with the inclusion of clustered objects in the object layer, which corresponds to multiple sensors in a single object, like a connected car having multiple sensors and electronic control units. Clustered objects notion is claimed to be significant for AC purposes by providing a first step filtering such that devices in the same cluster are checked if they can communicate with each other. Besides providing storage and computation capabilities, the cloud layer is suggested to be used for defining security policies to enable authorized communication of the connected vehicles. They present two use cases, one for single and one for multiple cloud systems, to demonstrate authorization requirements in the IoV.

In [61], we propose a risk-aware AC framework for CPSs that integrates the cyber–physical AC (CPAC) model and the generalized action generation model (GAGM). CPAC is proposed based on a set of AC requirements derived for CPSs. It extends the core ABAC model explained in Section 1.4.1 with components related to CPSs and interaction between those components. Figure 1.2 shows the structure of the CPAC model. There are five finite sets of elements in the CPAC: Agn, RS, Act, E and MO. Agn and RS represent the set of agents and resources, respectively. They consist of three types of elements, which are human, physical and cyber, so $Agn = HA \cup PA \cup CA$ and $RS = HR \cup PR \cup CR$. If we consider a smart traffic management system as an example, drivers are human agents, roads or road monitoring equipment are physical agents and sensors are cyber agents. Act denotes the set of actions that the agents can take on resources. It consists of two types of elements: object-oriented actions (OOA) and cyber–physical actions (CPA). OOA is as defined in the core ABAC, whereas CPA includes actions based on the application domain, like infusing a medication to a patient. E is a finite set for representing the environmental context including the physical environment and the time contexts. MO denotes a finite set of operational modes under which a CPS can operate. MO consists of emergency mode and nonemergency modes such as active, passive and autonomous modes [62]. Cyber–physical interaction or feedback loop mechanism, which are explained in Section 1.2.3.2, are reflected in the CPAC as the *role exchange* between agent and resource elements. To give an example, a sensor is an agent when it reads the glucose level of

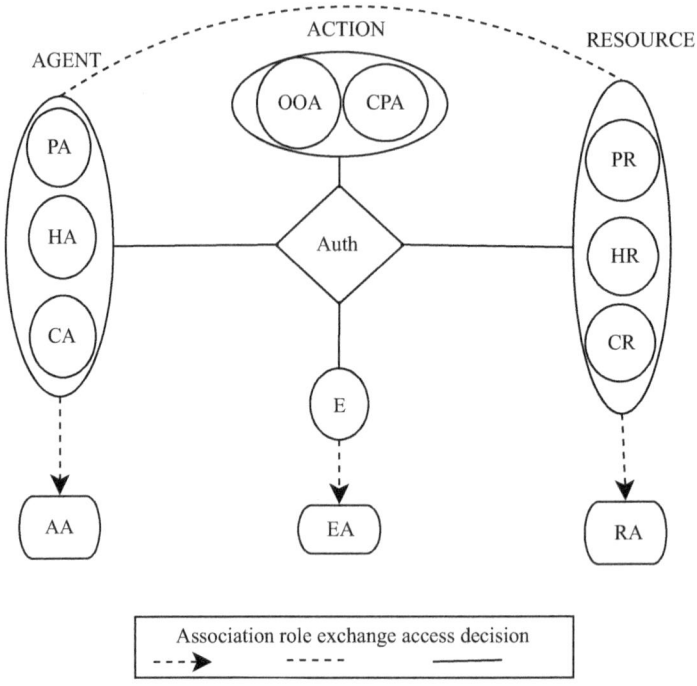

Figure 1.2   Structure of CPAC model

a patient and a resource when it sends the reading to the smart decision entity in an MCPS.

CPAC comprises four functions, which are *Attr, AttrType, Range* and *Auth* as in the core ABAC model. *Attr* is the finite set of attribute functions as explained earlier. In CPAC, it consists of a set of attribute functions for agents, resources and environment context, so *Attr = AA ∪ RA ∪ EA*. *AttrType* is the function which maps an attribute to its type, so *AttrType : Attr → {atomic, set}*. *Range* is the function that returns domain of an *attr ∈ Attr*, which consists of possible values *attr* can take. Therefore, if *AttrType(attr) = atomic*, then *attr(elem)* returns a single value from *Range(attr)*, where *elem ∈ Agn ∪ RS ∪ E*. However, if *AttrType(attr) = set*, then *attr(elem)* returns a set of values from the power set of *Range(attr)*, i.e. $2^{Range(attr)}$. *Auth* is the finite set of authorization policy rules such that ∀*auth ∈ Auth* : *Agn* × *Perm* × *MO* → *{permit, deny}*. This means that an *auth ∈ Auth* evaluates a *perm ∈ Perm* requested by an *agn ∈ Agn* under an operational mode *mo ∈ MO*, and returns a decision as permit or deny, where *perm =< act, rs >*, ∀*act ∈ Act* ∧ ∀*rs ∈ RS*.

GAGM is utilized to support the enforcement of AC policies. GAGM is a formally generalized version of the Action Generation Model (AGM) [63] to different operational modes of a CPS. GAGM is capable of handling nonemergency modes, such as active, passive and autonomous, whereas the AGM is only used for emergency mode and criticality management. We explain AGM briefly here. Next, we will explain GAGM of our proposed framework in [61]. AGM is proposed to manage active criticalities in CPSs in case of emergencies. Managing criticalities in CPSs is important [64], because they move the system to an unstable state if the required actions are not taken immediately in a predefined time interval [63,64]. One of the most prominent features of the AGM is that it proactively finds the authorization privileges needed by users in case of an emergency, without requiring them to submit access requests. To do so, the AGM calculates optimal path of actions to reach the normal state from a given critical state. Hence, the system states in the AGM are of two types as normal and critical. They are represented as nodes by the AGM. Links between the nodes are of two types as critical links (CLs) and response links (RLs). When the system is in the normal state, it enters into a critical state with the occurrence of a criticality. This is represented as a CL in the AGM. In a critical state, a set of response actions are taken to reach the normal state. This is represented as an RL in the AGM. Both types of links are labeled with probability values that reflect the probability of the occurrence of a criticality for a CL and the probability of successfully executing the response action for an RL. When all active critical events are solved, the system turns into the normal state. A difference between the GAGM and the AGM is the inclusion of risk in optimal action path calculations. The AGM only uses probability values of CLs and RLs in the calculation of optimal action paths. However, the risk of both critical events and unsuccessful execution of response actions are crucial for action planning in CPSs.

Here, we present a brief overview of the GAGM model. GAGM is based on the probabilistic finite automaton and represented as a tuple: $<G, Ev, Act, MO, \Theta, RI>$, where Act and MO are the sets as defined in the CPAC model. G is the set of graphs generated for different operational modes of a CPS. Each graph consists of a set of vertices, V_{m_i}, and a set of edges, E_{m_i}. Some of those vertices can be final states, $F \subseteq V$ and some of them can be initial states, $I \subseteq V$. An edge is labeled with an event, $ev \in Ev$, and its probability, $p_{ev} \in \Theta$, where Θ is a transition probability function. An event can be a critical event, $cEv \in CEv$, or an action, $act \in Act$. RI is the set of risk values associated with vertices in a graph. Figure 1.3 shows a sample GAGM to demonstrate the relation between graph of emergency mode and graphs of nonemergency modes, as presented in [61]. In this figure, G_{m_e} is connected to other graphs by normal states N_3, N_8, N_{11}. While calculating optimal action paths, or paths with minimum risk in the GAGM, the Dijkstra's algorithm [65] is used, because the problem suits to well known shortest path problem and the edge weights are nonnegative. Optimal paths from each initial state to a final state for all graphs are calculated in an offline manner and stored by the GAGM module of the proposed architecture. These stored data are later used in real-time operation of the AC system to retrieve the required actions to be taken in the current system state and the attributes of eligible agents who can perform these actions.

1.4.3 AC models and frameworks for secure interoperation

We discuss secure interoperation (or inter-domain AC) in the following section briefly. Next, we present various approaches for secure interoperation.

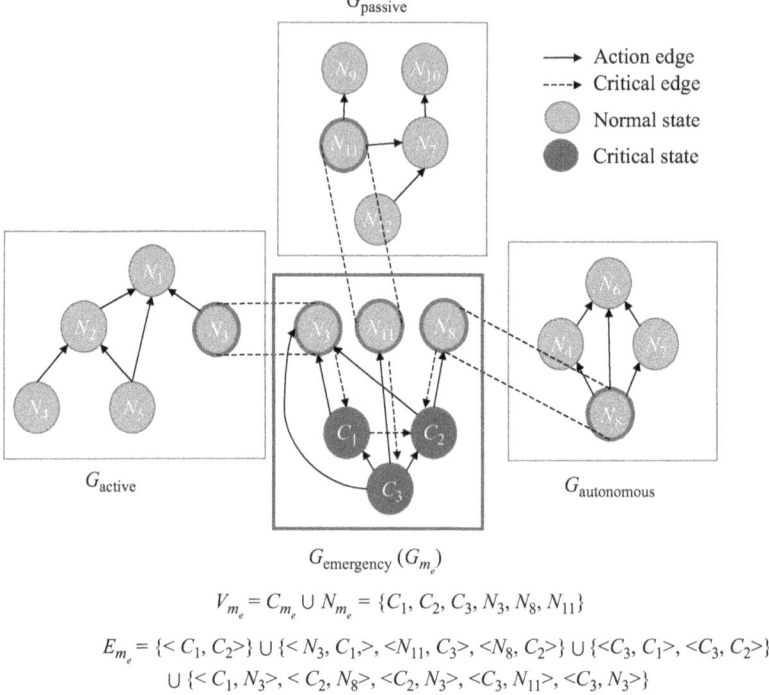

$V_{m_e} = C_{m_e} \cup N_{m_e} = \{C_1, C_2, C_3, N_3, N_8, N_{11}\}$

$E_{m_e} = \{<C_1, C_2>\} \cup \{<N_3, C_1>, <N_{11}, C_3>, <N_8, C_2>\} \cup \{<C_3, C_1>, <C_3, C_2>\}$
$\cup \{<C_1, N_3>, <C_2, N_8>, <C_2, N_3>, <C_3, N_{11}>, <C_3, N_3>\}$

Figure 1.3   An example GAGM representation from [61]

1.4.3.1   Basics of secure interoperation

We have already mentioned about inter-domain AC in Section 1.3.3. A *domain* can be considered as a computing environment including users and resources subject to the same set of AC policies. Accordingly, a *multi-domain environment* consists of several individual domains interacting to or collaborating with each other and having their own authorization requirements and AC policies [66]. Secure interoperation of AC policies in a multi-domain environment requires the analysis of the AC policies of participating domains in the cooperation. It also requires the construction of an *interoperation policy* emerging from the cooperation to enable legacy systems to share resources in a secure way. One should envision an interoperation policy as a contract signed by the organizations participating in the collaboration and governing the interoperation among them [67]. It is used to define the rules of cross-domain authorization to decide whether a user demanding access to the resources in the provider domain can acquire the required permissions, by checking his privileges [37]. As a result, the secure interoperation problem transforms into the interoperation of AC policies of the interacting organizations [67].

Figure 1.4 demonstrates an abstract view of the secure interoperation of multiple domains from an AC perspective. In this figure, RBAC is assumed to be used by all the participating domains. In each domain, there is a role hierarchy indicating that a senior

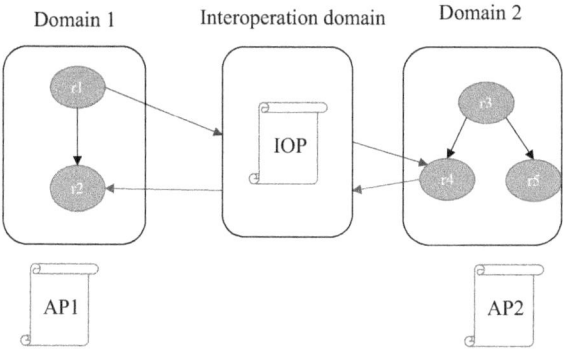

Figure 1.4   An abstract view of secure interoperation under RBAC

role may obtain the permissions assigned to their junior roles. Directions of an arc shows the direction of the hierarchy. For example, Role 3 can acquire the permissions assigned to Roles 4 and 5 in Domain 2. Each domain has its own AC policy, shown as AP1 and AP2 for Domains 1 and 2, respectively. An interoperation policy IOP is formed within the interoperation domain, governing the interoperation between the two domains. Hence, Domains 1 and 2 can share resources without revealing their policies to each other. In case of RBAC, interoperation policy composition corresponds to *role mapping* between individual domains, based on the resource and information sharing policies defined by individual domains, and permissions assigned to those roles [53]. Role mapping provides a user of a domain with a specific role to acquire the permissions of the mapped role in another domain. In Figure 1.4, Role 1 in Domain 1 can acquire the permissions assigned to Role 4 in Domain 2.

When multiple parties join into a collaboration and want to interoperate in a secure way, there is a high probability that each individual domain has its own specification language to state its AC policies [68]. This can cause several policy inconsistency problems in the multi-domain environment such as conflicts in naming, in role hierarchies and in the policy constraints [53]. To give an example, representing the same role by different names in different domains, such as junior nurse and intern nurse, would cause a name and role heterogeneity. This type of conflicts can be resolved by applying database schema integration methodologies, or ontology-based approaches [53,68]. Secure interoperation of AC policies in a multi-domain environment requires to meet *secure interoperation principles*, namely *security* and *autonomy* principles [69]. The security principle requires that the access denied in one domain should still be denied in that domain after an interoperation policy is formed. Autonomy principle requires that the access allowed in one domain should still be allowed in that domain after an interoperation policy is formed.

There are several approaches for providing secure interoperation among multiple domains from an AC perspective. One approach is measuring the similarity of the authorization policies of collaborating parties. Proposed solutions in this category [70,71] aim at *(i)* finding the parties which have similar AC policies to maintain the same level of security when they collaborate and *(ii)* maximizing the number of resources to be accessed by the users residing in different domains. Another approach is the use of policy combination or policy integration methods [72,73]. In addition, policy mapping techniques can be applied which are known as role mapping in case of the RBAC model. Although there are several works that provide frameworks for constructing secure interoperation RBAC policies by using role mapping, very little effort has been put on the secure interoperation frameworks for ABAC policies yet. This is caused by the fact that ABAC is still in its early development phase.

1.4.3.2   Methods and frameworks for secure interoperation

Previously proposed secure interoperation frameworks aim at providing solutions which form a "global policy" (or interoperation policy) from the policies of individual collaborating domains so that accesses in inter-domain (or multi-domain) environment are governed, as explained in Section 1.4.3.1. Shafiq *et al.* in [53] propose such a framework for integrating RBAC policies of participating domains to obtain a global policy. They use a graph-based formalism for representing the RBAC policies of local domains and the interoperation domain. Users, roles, and permissions are represented as nodes. Relationships between those elements (such as user to role assignment, permission to role assignment, inheritance hierarchy (I-hierarchy) or activation hierarchy (A-hierarchy), SoD constraint) are represented as edges. Policy integration requirements are defined as: the preservation of *(i)* permissions provided to the users in individual domains, *(ii)* existing elements (the set of users, permissions and roles) in individual policies and *(iii)* relationships between those elements in the multi-domain policy, *(iv)* obtaining the same multi-domain policy independent of the order of the given input policies and *(v)* satisfying security requirements (role assignment and SoD constraints). The proposed framework first composes an interoperation policy via a role mapping mechanism. However, before composing this policy, individual domains need to decide their information sharing policy with external domains because a domain may not want to reveal all the data (or shareable objects) it has, as they are. Therefore, for each shareable object, a domain needs to decide with which external domains to share objects, which parts of the data to sanitize, access mode of the objects to share and so on. Another issue considered before the multi-domain policy composition is to solve several heterogeneity issues that are naming conflicts, role hierarchy conflicts and policy constraint conflicts as mentioned in Section 1.4.3.1.

The next step performed by the proposed framework is the resolution of conflicts in the generated interoperation policy. The reason is that the interoperation policy may be inconsistent in terms of the security requirements of local domains. Three security requirements are the focus of the framework: role assignment, role-specific SoD and user-specific SoD. Authors define role-assignment constraint as the "security principle," one of the secure interoperation principles discussed earlier in Section 1.4.3.1. Therefore, based on this constraint, after the interoperation policy formed, a user should not be able to access a local role "r" or its senior roles, if it is originally not allowed in the local domain. Role-specific SoD constraint requires that a user is not allowed to activate conflicting roles in the same session or concurrent sessions. User-specific SoD constraint indicates that conflicting users for a role are not allowed to activate this role in the same session or concurrent sessions. Authors propose a conflict resolution algorithm based on the integer programming approach that removes some of the role mappings generated for the interoperation policy. This algorithm converts role-assignment, role-specific SoD and user-specific SoD constraints of RBAC policies to integer programming constraints. Then with an objective function which maximizes the number of accesses, the formulated IP problem is solved to get the final interoperation policy.

Another secure interoperation framework for RBAC policies is proposed in [74] to support information sharing in a multi-domain environment. This framework addresses several limitations of the previously proposed frameworks. First, policies of individual domains are not modified as opposed to previous frameworks. In addition, policy constraints of individual domains, namely temporal constraints, SSoD and DSoD constraints, are tackled when interoperation policies are constructed between external and internal domains. Authors define an external domain as the one which submits access requests and an internal domain as the one which provides resources in response to the requests by external domains. Accordingly, an interoperation policy is defined as an intermediate level between an external and an internal domain that manages accesses to the resources of the internal domain by the external domain users. Secure interoperation is enabled by extending the policy of an internal domain with the interoperation policy. The resulting policy is called as *interoperation augmented policy (IAP)*. An interoperation policy is generated as a result of a set of *interoperation queries*, which are submitted by security administrators of internal domains. For each external domain, an interoperation policy is generated. An interoperation query is represented as $Q = < r, PS, \tau >$, where r denotes the role of the user in an external domain who requests the access, PS is a set of permissions needed to access resources and r is a periodic time expression used to denote the time which the access is needed. An algorithm is proposed for constructing the interoperation policy, which consists of two steps: role selection (role mapping) and interoperation policy construction. In the first step, a set of roles in the internal domain are selected that provide the requested permissions in the interoperation query and have the maximum time coverage for interoperation. The role selection process is formulated as an optimization problem with backtracking approach. In the second step, the proposed framework constructs artificial filter roles for the interoperation policy. Filter roles are used for providing the least privilege principle. They have upper bound set of granted permissions for the requesting user of the external domain, so excessive permissions are not granted.

Cruz *et al.* [75] propose an interoperation framework using RBAC as the underlying AC model. They utilize from the Web Ontology Language (OWL) for representing both individual and interoperation policies. Their approach is distinguished from prior approaches in terms of selectively granting permissions to a resource requester who is outside of the provider domain. Other approaches concern to provide all the permissions to requesting users as long as they activate the associated roles in the provider domain. In addition, they consider *interoperation constraints* to be enforced on role mappings in case individual domains want to limit the access on their resources by limiting global access rights. Baliosian and Cavalli [67] automate the process of secure interoperation policy construction using a form of finite state machines: finite state transducers. Accordingly, they define a policy rule as a transition from one state to another state of the entities in the system. As the AC model, they use organization-based AC (OrBAC) model. They also provide a methodology for mapping the names of the entities in participating domains which are syntactically different but semantically the same (like mapping physician in domain 1 to the health-care professional in domain 2). The mapping process is handled manually by security administrators.

Different than single-domain AC models, the models developed for collaboration among multiple domains require to meet secure interoperation principles. Gouglidis *et al.* [76] tackle those principles of autonomy and security and propose formal definitions for them. They present a methodology to verify secure interoperation policies. They use domRBAC, which is an extended version of the RBAC with multi-domain collaboration notion, as the underlying AC model. In order to verify interoperation AC policies, they propose the use of model checking technique for detecting violations against security and autonomy principles. Specifically, they look at cyclic inheritance, privilege escalation and SoD properties to detect violations. El Maarabani *et al.* [77] develop a framework to check the validity of secure interoperation policies. They propose formal validity testing of interoperation policies by using finite automata. In their solution, OrBAC is the underlying AC model for secure interoperation. They also propose the integration of policy rules in each domain for the secure interoperation between internal and external domains.

The secure interoperation frameworks discussed so far build upon existing AC models such as the RBAC and OrBAC. Different than those frameworks, Jung and Joshi [78] propose the community-centric property-based AC (CPBAC) model to enable secure and effective collaboration in Online Social Networks (OSNs). Since OSNs also support smart cities by means of data acquisition purposes [2] and enable collaboration and information sharing among users, it is important to overview such an AC model. CPBAC is a property-based model, built on the requirements derived for AC and secure cooperation in OSNs. Those requirements include consideration of social relationships, fine granularity of control on personal information, individualized policies for users, sticky policy with the users' data, interoperability, anonymous cooperation and time-based control. CPBAC is an extension of the previously proposed community-centric role interaction-based AC (CRiBAC) model in OSNs domain. CRiBAC is based on the role-based approach and proposed for enabling secure collaboration in MASs. The main components of CRiBAC are agents, community-based components (society, community, community role), roles, object-oriented and interaction permissions. Interaction permissions are of three types: resource-oriented permissions, role-oriented permissions and task-oriented permissions. In this model, agents participate in communities with community roles. Communities are generated dynamically based on the occasions to achieve a common goal by the agents and destroyed after the goal is accomplished. They are associated with context information such as the community type, the time of community creation and the member information. *CRA* represents the policy for assignment of agents to community roles, based on the context and tasks of agents.

Unlike CRiBAC, CPBAC includes a "user" component in place of an "agent" component. In addition, each entity is associated with a set of properties in CPBAC, which makes it a property-based AC model. Properties of users are defined as contexts (age, gender, job etc.), tasks, user resources and user policies. Properties of communities are contexts, tasks, community resources, cooperation process, community policies (include AC policy, recruiting policy for deciding eligible users and SoD policy) and community members. A society in CPBAC is an OSN entity which supports secure cooperation. The properties of society comprise contexts (like number of users), tasks, society resources and society policies. Permissions in CPBAC have five categories, one of which is a regular permission as in RBAC (<object, operation> pair) and the remaining four are based on object properties (context-oriented, task-oriented, resource-oriented, policy-oriented). CPBAC supports eligibility constraints (on users to join communities), in addition to the SoD and cardinality constraints as in the RBAC model. An administrative model is also proposed for CPBAC (ACPBAC) in [78] that is for administrating users and privileges efficiently. It is a decentralized administrative model that involves three types of administrators (society manager, community manager and user) and distribute administrative responsibilities to them. Administrative functions are basically for assigning and deassigning users to permissions and to communities/society. Authors build a prototype application, named as Secure Cooperative OSN (SeCON), to demonstrate that the CPBAC is feasible and how it can be implemented in a real-world emergency situation, such as finding a lost child in an OSN.

1.4.4 Trust-based AC models and frameworks

Given the importance of trust management mechanisms in smart cities as explained in Section 1.3.5, in this section, we overview the AC frameworks incorporating trust.

Bernabe *et al.* [79] propose a trust-aware AC system for the IoT (TACIoT). TACIoT incorporates an AC model and a trust model, which are designed for special requirements of the IoT environments and devices, namely being flexible and lightweight to deal with the heterogeneity and ubiquity of IoT devices. TACIoT system is aimed as a subcomponent of the IoT security framework previously proposed by the authors, which is based on the IoT-ARM architectural model [56] mentioned earlier in Section 1.4.2. Components of the IoT security framework to be used by the TACIoT system are the authorization manager, the context manager and the trust manager. Context manager provides necessary context information to authorization and trust manager components to make decisions, such as the time of the day and the battery level of an IoT device. Authorization manager makes authorization decisions using its policy decision point (PDP) subcomponent, based on XACML formatted ABAC policies. Owing to *obligations* part in XACML policies, the PDP can state the necessary actions to be taken by the policy enforcement point (PEP) implemented on a target device. In this way, trust conditions to authorizations are enforced during token validation process on the target device. Authorization manager has also a token manager subcomponent, which generates access tokens for permit decisions and signs them using Elliptic Curve Cryptography scheme. Trust manager is in charge of computing and providing the trust values of the devices requesting accesses. It can be deployed on the target device if it is a non-constrained device such as a smartphone, or on another device otherwise. Authorization manager is not deployed on target devices since it needs to store authorization policies, and this will not be feasible for constrained IoT devices.

Authors in [79] discuss that the majority of the available AC mechanisms for the IoT are based on a centralized approach due to resource-constrained devices. They argue that a centralized approach to AC in the IoT has its own drawbacks, such as requiring frequent message exchanges between constrained devices and the central entity, which will be another source of resource consumption. Therefore, they adopt a distributed AC mechanism, distributed capability-based AC (DCapBAC) [57] that is previously proposed by them. DCapBAC is proposed based on the constraints of the IoT devices. As explained previously, in DCapBAC, and so in TACIoT, tokens are used as capabilities by subjects requesting accesses. Access requests are handled on the target device by validating access tokens of subjects. It is done by verifying the signature of the token and checking the trust value of the subject against the trust threshold placed in the token. In this way, once a subject obtains a token, it can be used until being expired or revoked and there is no need to evaluate access requests against AC policies each time.

The proposed trust model for the TACIoT [79] is novel compared to the previous approaches in terms of being multidimensional for more accurate trust calculations. It considers the QoS, security, reputation and social relationships of devices as trust properties. QoS property consists of four sub-properties: percentage of successful interactions, availability, throughput and delay. Similarly, security property has four sub-properties: supported authentication and authorization mechanisms by the device, employed communication protocol (like IPSec and TLS), the network scope (being in the same local network or not) and the intelligence capacity of a device (probability of behaving maliciously is assumed to have relation with powerful computing resources). These security properties are obtained in the real time via the Context Manager component. Reputation property denotes the recommendation about a device requesting access provided by others. It is an indirect trust measure, calculated by multiplying the opinion of a device j on a requesting device i and the credibility of device j. To prevent recommendations from dishonest devices, they use reputation only if the difference between the reputation and the direct trust observation by the device is below a certain threshold. Social relationship property is about Social IoT notion, where IoT devices are assumed to have social relationships with each other and are assigned to *bubbles* and *communities*. The trust between two devices changes based on the closeness of their relationship, i.e. the type of the bubble, like personal, family, or office bubble. All sub properties are aggregated by using weighted sum and then normalized into [0, 1] interval. Finally, normalized main trust properties are merged using the fuzzy logic owing to its strength on managing uncertain, aggregated and subjective measures. A fuzzy control system is deployed in the trust management component that takes continuous trust properties from the interval [0, 1] as input and produces the outputs represented as fuzzy linguistic variables, which are distrust, untrust, trust and high trust. Authors implement the TACIoT to see its feasibility and performance both on constrained IoT devices, like sensors, and on non-constrained devices, like smartphones.

Pustchi and Sandhu [54] propose a multi-tenant attribute-based AC (MTABAC) model by extending the core ABAC model explained in Section 1.4.1 with cross-tenant trust concept. MTABAC aims at enabling collaboration among the tenants of a single cloud system, specifically for IaaS (Infrastructure-as-a-Service), where AC is based on attributes and trust relations. In a cloud system, customers are represented as tenants, such as an organization, a department, or individual users. Resources of customers are organized in separated containers, which are virtual computing platforms, to provide better security and privacy. An example scenario for multi-tenant collaboration in a cloud system is presented in the paper. Sample tenants are software development, software testing and system engineering departments of an organization. In this scenario, and in MTABAC in general, it is aimed that the users of a tenant are able to access the resources of another tenant based on cross-tenant trust attribute value assignments.

Compared to the core ABAC, MTABAC includes an additional entity: a set of tenants (T). Users and objects belong to tenants and two different "one-to-many atomic attribute functions" are defined to represent these relationships: *userOwner* and *objOwner*, respectively. They also define two meta attributes, *uattOwner* and *oattOwner*, as user and object attributes are owned by tenants. Another reason for such a design is that each tenant manages its own attributes, i.e. assigns values to them, add or delete them. In MTABAC, the trust is modeled by means of an attribute of tenants, *trustedTenants* : $(t : T) \rightarrow 2^T$. It is represented as $T_A \trianglelefteq T_B$, which means tenant A is a *trustor* tenant and trusts tenant B as a *trustee* tenant, where $T_B \in trustedTenants(T_A)$. By means of meta attributes and tenant–trust relationship, a trustor domain has control over trust relations. A trustee tenant can assign attribute values to the users of a trustor tenant when they request access to the resources in the trustee tenant domain. However, this is not the case for object attributes, i.e. a trustee tenant cannot assign attribute values to the objects in the trustor tenant domain. Authors also discuss about multi-tenant role-based AC (MTRBAC) model, the version which is distilled from several previously proposed approaches [80–82], and show that MTABAC can be configured as MTRBAC. MTRBAC is based on the RBAC approach that is explained earlier in Section 1.4.1. It has a set of tenants, T, as an additional component to the RBAC. Similar to MTABAC, MTRBAC includes ownership relations between tenants and other entities: user ownership (UO), role ownership (RO) and object ownership (OO). In MTRBAC, trust relationships are used by a trustee tenant to assign roles to the users in a trustor tenant domain when they want to access resources in the trustee domain. As a result, MTABAC is configured to behave as the MTRBAC by the use of a role attribute function and by representing RO as the aforementioned meta attributes of MTABAC.

In [81], OpenStack AC domain trust (OSAC-DT) model is proposed for the OpenStack that is an open source cloud IaaS platform. First, the existing AC model applied by the OpenStack is formalized by utilizing OpenStack Identity API. Authors name this formal model as OpenStack AC (OSAC) model. OSAC is based on the standardized RBAC model [83] and extends it for supporting multiple tenants. The components of the OSAC are users (U), roles (R), groups (G), domains (D), projects (P), services (S), object types (OT), operations (O) and tokens (T). Users belong to domains and are managed by administrators of domains. Domain corresponds to the tenant concept explained earlier. It is considered as an administrative boundary of users, groups and projects. An administrator of a domain can access to the information about all the users of a domain and assign them to the roles in that domain. A group contains users of a domain and belongs to a single domain. Each project is owned by a single domain and defined as a container which holds resources of the domain and manages cloud services. Services are basically applications and resources provided by cloud systems, such as networking, computation, image and so on. An object type and an operation together form a permission, where the object type represents a cloud resource category like virtual machine and the operation is an access mode as explained earlier. Tokens are used by users to access the resources. They are provided to authenticated users by the identity service and have an expiration date. A token includes authorization-related information that is encrypted by the public key infrastructure to protect it against modification, such as the user identity requesting access, the project to be accessed, roles of the user in the project and services in the project. The functions defined in OSAC are user_owner UO, group_owner GO, project_owner PO, user_tokens, token_project and token_roles. UO, GO and PO denote mappings of users, groups and projects to domains as discussed earlier. User_tokens, token_project and token_roles are for mapping users, projects and roles to tokens. There are three assignments in the OSAC, which are user assignment, group assignment and permission assignment, where users and groups are assigned to (project, role) pairs and permissions are assigned to roles. Moreover, an administrative model for the OSAC (AOSAC) is proposed by the authors that involves cloud admin, domain admin and project admin roles for the administration of authorizations, from the highest to the lowest level of authority, respectively. For example, a project admin deals with the assignment of users and groups to (project, role) pairs.

Authors in [81] extend the OSAC with domain trust relationship for enabling secure cross-domain authorization. To motivate the need of their extension, they present a sample scenario for DevOps, which corresponds to a software development strategy for quick production of the software via collaboration among different departments. In this scenario, there are two collaborating domains: production and development, where the production domain comprises projects related to the daily business and the development domain provides sandboxing environment for developers and testers. Characteristics of domain trust relations in the OSAC-DT model are two-party, unilateral, unidirectional and non-transitive. Two-party trust means the trust relationship is established between two domains, compared to federated trust where the trust is established among multiple domains. Unilateral trust is about the initiation of the trust relationship, where trusting party establishes the relationship without confirmation from the other party, compared to the bilateral trust. Unidirectional trust entitles only one domain for the actions enabled by trust relationship, whereas bidirectional treats both the trusting and trusted parties equal from this perspective. Non-transitive trust does not allow the transfer of the trust from an existing trust relationship between two domains to other domains. The trust relation is a many-to-many mapping between domains and is represented as follows: $A \trianglelefteq B$, A is being the trustor and B is being the trustee. Similarly, trustor domain has control on the trust relationship and the trustee handles role assignments for cross-domain AC. Authors categorize trust relationships into three: Type-α ($\trianglelefteq_\alpha$), Type-β ($\trianglelefteq_\beta$) and Type-γ ($\trianglelefteq_\gamma$). In Type-α, the user information of the trusted domain should be exposed to the *trusting* domain for the assignment of users to (project, role) pairs, whereas in Type-β the user information of the *trustor* domain needs to be exposed. Type-γ discloses the project information of the trustor domain to the trustee domain for the same purpose. Finally, experiments are performed on a prototype implementation of the OSAC-DT, which extends the OpenStack with trust relations. Results show that the OSAC-DT brings minimum overhead onto authorization, so it can be used for cross-domain AC purposes for the OpenStack platform.

1.4.5 AC models and frameworks for insider threats

Nasr *et al.* in [84] propose an alarm and trust-based access management system (ATAMS) for protecting SCADA systems against deontological threats that is a type of insider threat performed against SCADA systems. The purpose of the ATAMS is to minimize the access of an attacker to critical substations. An integrity policy is used to control access of operators to substations. For this purpose, substations are labeled with integrity levels and operators are assigned with trust values. Initially, operators are assigned with permissions to access substations in their area of responsibility (AoR) by a supervisor using the RBAC model. Later, when a deontological threat by an operator is detected, its trust value is decreased and the access of the operator to a remote substation/field device in his/her AoR is determined by comparing his/her trust level to the integrity level of the substation. They compute the integrity level of a substation using "substation criticality" as a proxy. The criticality of a substation is computed based on substation properties, such as substation capacity and the number of transformers. Trust of an operator depends on its current and historical performance to resolve alarms over an acceptable time that are raised by field devices in case of faults. Performance of an operator is calculated as a weighted average of the total number of resolved alarms where the weight is the severity of an alarm. Current trust value of an operator is computed by combining historical trust values using exponentially weighted moving average. They propose an anomaly detection method for SCADA systems based on the severity of alarms and their resolution time using a statistical quality control technique, Shewart control charts. Severity of an alarm is computed based on the alarm type and the integrity level of the substation by which it is generated. If an operator cannot resolve an alarm within its resolution time, a "fault condition" value is computed for a deontological threat based on the severity of the alarm. If this value falls outside the control limits, then the case is labeled as an anomaly. They evaluate the effectiveness of the ATAMS under two scenarios: intentional and unintentional deontological attack. They extract real data from the SCADA in Iran power system that contains events/alarms from substations and simulate deontological attack using parameters obtained from the real data, such as alarm distribution of each substation and distribution of time spent to clear alarms.

Risk-aware approach to AC has been considered as a fundamental mechanism to tackle insider threats. In this direction, Bijon *et al.* [85] propose an adaptive risk-aware RBAC framework. They present a formal model for adaptive risk-aware RBAC by extending the NIST core RBAC model [83]. They specify risk-aware components of the RBAC as user-role assignment, permission-role assignment, sessions, role hierarchy and constraints. Even if a single permission is not dangerous on its own, when several permissions are combined together and assigned to a role, the role can be risky to an organization. Likewise, the combination of roles assigned to a user may result in severe risks. As explained earlier in Section 1.4.1, users activate the roles assigned to them in sessions to obtain permissions required to complete tasks. Activation of certain roles in a session affects the risk exposure of the system. In addition, the risk associated with a session depends on the context, such as the time and location of the access. An AC framework built on risk-aware RBAC approach includes a risk engine which is responsible for risk management activities such as specification and verification of RBAC constraints and estimation and adjustment of risk thresholds. Authors categorize risk-aware approaches to RBAC into two as *traditional* and *quantified*. The main difference between traditional and quantified risk mitigation approaches is the existence of an explicit risk value. In traditional approach, administrators decide risky operations and policies to enforce constraints on these operations, whereas in a quantified approach, risk values and thresholds are calculated dynamically by the system with respect to processes and formulas. Their proposed RBAC framework is based on the quantified adaptive risk approach.

In a quantified risk-aware RBAC session, the *risk value* of the roles activated by a user in a session is compared to the estimated *risk threshold* of the session. The risk value of a session is the sum of the risk values of individual roles activated by the user. The risk value of a role is calculated as the sum of the risk values of individual permissions assigned to the role. The risk threshold of a session is estimated based on *risk factors*, such as purpose, connection and access history. If the risk of a session is higher than the risk threshold, then some of the activated roles needs to be deactivated. This property makes quantified risk-aware RBAC a dynamic approach, because the access capability of a user changes with respect to the risk estimation of the current session. Different than this nonadaptive quantified risk mitigation approach, an adaptive type of quantified risk mitigation approach requires continuous monitoring of behaviors in a system. Monitoring and anomaly detection tasks are the duty of a risk engine. In case an abnormal event occurs, the risk engine adjusts the risk threshold of the user session accordingly, by using role deactivation mechanism. In addition to the monitoring of anomalous behaviors and adjustment of the risk threshold for a session, a similar mechanism is required for user and role components of a risk-aware RBAC. To be more precise, in case the risk threshold of a user or a role is exceeded, the roles assigned to the user and the permissions assigned to the role should be revoked, respectively.

In the proposed adaptive quantified risk-aware RBAC framework in [85], users create sessions and then activate roles assigned to them as in the core RBAC. Deactivation of the roles is an iterative process carried out automatically by the proposed framework. When the risk value of the session exceeds the risk threshold, a list of activated roles to be deactivated is decided by the system and displayed to the user. Then the user selects the roles to deactivate and this process repeats until the risk value drops below the threshold. The formal model presented in the paper for adaptive quantified risk-aware RBAC includes the following basic functions: *assigned_risk*, *risk_threshold* and *session_risk*. Function *assigned_risk* maps a permission to a risk value whereas *risk_threshold* and *session_risk* are mappings of sessions to the maximum risk value and the current risk value of a session, respectively. Besides these basic functions, there are administrative and system functions in the proposed model, which are *AssignRisk, RoleRisk, Create-Session, AddActiveRole, Deactivation, SActivityMonitor* and *SADeactivation*. A user can only invoke *CreateSession* and *AddActiveRole* functions. Remaining functions are invoked by the system. *CreateSession* function makes use of a function to generate an estimation for the risk threshold of the current session. The tasks performed by *AddActiveRole* and *Deactivation* functions are the activation of roles in a session by a user and the deactivation of roles by the system. *SActivityMonitor* function is called consistently during a session. It monitors user activities via *SADetection* function and it triggers a function to reevaluate the risk threshold of the current session if an anomaly is detected. *SADeactivation* function is also a part of *SActivityMonitor* function, which takes care of the list of risky roles to be deactivated.

Baracaldo and Joshi [40] propose a trust and risk-aware AC framework for preventing insider threats. They extend RBAC with trust and risk. The proposed AC framework is adaptive in terms of user behaviors such that it removes access privileges when the trust value of a user drops below a threshold. The trust threshold is required for a user to activate a set of roles in a session, among the ones assigned to the user. They address the role activation problem of the RBAC and propose a role activation algorithm based on the proposed trust threshold computations and RBAC constraints. The constraints taken into account are SSoD, DSoD and cardinality constraints. SSoD and DSoD are defined as explained earlier in Section 1.4.1. Cardinality constraints they consider are of two types: (*i*) a role can be assigned at most to a specified number of users and (*ii*) a role can be activated at most by a specified number of users. Their extended RBAC model also supports hybrid role hierarchies, which is a combination of role inheritance and role activation hierarchies as explained earlier in Section 1.4.1. The reason to include hybrid hierarchy is that it allows to enforce DSoD constraints. They also introduce the concept of *inference threat*, which corresponds to the case that a user can infer extra permissions that is not authorized to him originally in the policy. They formulate a Colored Petri Net (CP-Net) to model the history of role activations by a user and to detect inferred unauthorized permissions. They provide a method for the calculation of the trust threshold, which is based on a risk assessment approach. The trust threshold involves both the risk of the requested permission under the context of the access, where the context can be the type of a service and the network connection, and the risk of inferred unauthorized permissions. The trust threshold is calculated as the ratio of the total risk occurring due to the activation of a set of permissions by a user, to the total risk of all the permissions in the system. Similar to the trust threshold and the risk of permissions, the trust value of a user depends on the current system and user context. The proposed role activation algorithm incorporates all the concepts discussed up to this point, i.e. SoD constraints, cardinality constraints, trust threshold, and aims to minimize the risk exposure of the organization while enforcing authorization policies. To be more precise, it evaluates an access request by checking whether (*i*) the requested roles for activation are assigned to the user, (*ii*) those roles respect to SoD and cardinality constraints and (*iii*) the trust value of the user is greater than or equal to the trust threshold required for roles to be activated, under the current context. If the conditions are met, the role activation algorithm finds an optimal set of roles to be activated, i.e. role set with minimum risk.

1.5   Future research directions and perspectives

As highlighted by Sicari *et al.* [41], there is a gap in the trust management literature for fully distributed and dynamic solutions for the IoT environments. When the vast number of the Things and scalability issues in conjunction with this are considered, the significance of such trust management schemes will be appreciated. The same issue applies to trust-based AC frameworks for the IoT and smart cities, since AC and identity management mechanisms are relied heavily upon to meet trust requirements [41]. We believe that by leveraging blockchain technology and its promise for decentralized trust, distributed AC approaches can be a promising direction for AC in smart cities. In addition to this, as we mentioned earlier, secure interoperation is essential for smart cities due to high probability that services are provided to citizens through interaction/interoperation among multiple parties in smart cities. Blockchain can pave the way for multi-domain AC to enable multiple organizations to collaborate and share information securely. One more future research direction would be devising AC mechanisms for attribute-based approach which aim to attain secure interoperation goal in smart cities. Even though there are promising proposals for secure interoperation of RBAC policies, research related to secure interoperation of ABAC policies is still in infancy.

1.6   Conclusions

Urban population in cities is rapidly increasing due to the growth in world population and fast urbanization. As a result of this, environmental problems have been growing at an unprecedented rate, such as city waste, greenhouse gases, resource and energy consumption. Smart cities are seen as solution to address these environmental issues such as by helping to prevent spread of contagious diseases. In addition to this, smart cities can improve quality of citizen's life via services such as the ones enabled by smart traffic management and smart health applications. Another area of improvement to cities provided by smart cities is efficient public administration and urban operations by cutting down extra costs involved in processes. To achieve these, smart city applications leverage various technologies. In this chapter, we have overviewed smart city applications, the IoT and CPSs as enabling technologies of smart cities, and the relation between the two. The technological constituents of the smart city infrastructure need to be integrated and they interact with each other for provisioning services to citizens. In addition, each enabling technology has its own unique characteristics. The interplay among smart city constituents and their unique characteristics pose unique security and privacy challenges. In this chapter, we have presented a high-level overview of the security and privacy challenges of smart cities and the unique characteristics of smart cities that make these challenges significant hurdles for the success of smart cities. Furthermore, we discussed some AC approaches from the literature which can be potential solutions to alleviate the security issues. The approaches we present include both the ones proposed specifically for smart cities and their enabling technologies, and the ones that address security risk, trust and secure interoperation issues. We observed that there is a wealth of AC models and frameworks built on role-based approach due to its maturity, whereas attribute-based AC domain needs more exploration. In this direction, we expect that secure interoperability of ABAC policies is a potential research direction.

Acknowledgment

This research work has been supported by the National Science Foundation grant DGE-1438809.

References

[1]   Mohanty SP, Choppali U, and Kougianos E. Everything you wanted to know about smart cities: The Internet of things is the backbone. IEEE Consumer Electronics Magazine. 2016;5(3):60–70.

[2]   Gharaibeh A, Salahuddin MA, Hussini SJ, *et al*. Smart cities: A survey on data management, security, and enabling technologies. IEEE Communications Surveys & Tutorials. 2017;19(4):2456–2501.

[3]   Complete Visual Networking Index (VNI) Forecast. Cisco; 2019 [updated 2019 Feb 18; cited 2019 May 10]. Available from: https://www.cisco.com/c/en/us/solutions/service-provider/visual-networking-index-vni/index.html.

[4]   Al-Turjman F, Ever E, and Zahmatkesh H. Small cells in the forthcoming 5G/IoT: Traffic modelling and deployment overview. IEEE Communications Surveys & Tutorials. 2018;21(1):28–65.

[5]   Al-Turjman F, and Baali I. Machine learning for wearable IoT-based applications: A survey. Transactions on Emerging Telecommunications Technologies. 2019; p. e3635. Available from: https://onlinelibrary.wiley.com/action/showCitFormats?doi=10.1002%2Fett.3635

[6]   Al-Turjman F, and Alturjman S. Multimed Tools Appl (2018). Available from: https://doi.org/10.1007/s11042-018-6288-7

[7]   Beltran V, Martinez J, and Skarmeta A. User-centric access control for efficient security in smart cities. In: Global Internet of Things Summit (GIoTS), 2017. IEEE; 2017. pp. 1–6.

[8]   DDoS Attack That Disrupted Internet Was Largest of Its Kind in History. Woolf, Nicky; 2016 [cited 2018 Mar 15]. Available from: https://www.theguardian.com/technology/2016/oct/26/ddos-attack-dyn-mirai-botnet.

[9]   Fernandez-Anez V. Stakeholders approach to smart cities: A survey on smart city definitions. In: International Conference on Smart Cities. Cham: Springer; 2016. pp. 157–167.

[10] Harrison C, Eckman B, Hamilton R, *et al*. Foundations for smarter cities. IBM Journal of Research and Development. 2010;54(4):1–16.

[11] Kondepudi S, Ramanarayanan V, Jain A, *et al*. Smart sustainable cities: An analysis of definitions. The ITU-T Focus Group for Smart Sustainable Cities; 2014 [updated 2014 Oct 18; cited 2020 Jan 08]. Available

from: http://www.itu.int/en/ITU-T/focusgroups/ssc/Documents/website/web-fg-ssc-0100-r9-definitions_technical_report.docx.

[12]   Zanella A, Bui N, Castellani A, *et al*. Internet of things for smart cities. IEEE Internet of Things journal. 2014;1(1):22–32.

[13]   Zhang K, Ni J, Yang K, *et al*. Security and privacy in smart city applications: Challenges and solutions. IEEE Communications Magazine. 2017;55(1): 122–129.

[14]   Arasteh H, Hosseinnezhad V, Loia V, *et al*. Iot-based smart cities: A survey. In: Environment and Electrical Engineering (EEEIC), 2016 IEEE 16th International Conference on. IEEE; 2016. pp. 1–6.

[15]   Djahel S, Doolan R, Muntean GM, *et al*. A communications-oriented perspective on traffic management systems for smart cities: Challenges and innovative approaches. IEEE Communications Surveys & Tutorials. 2015;17(1):125–151.

[16]   Cao QH, Khan I, Farahbakhsh R, *et al*. A trust model for data sharing in smart cities. In: Communications (ICC), 2016 IEEE International Conference on. IEEE; 2016. pp. 1–7.

[17]   A Role for NIST in Smart City Technologies. NIST; 2018 [updated 2018 Mar 20; cited 2018 Apr 12]. Available from: https://www.nist.gov/el/cyber-physical-systems/role-nist-smart-city-technologies.

[18]   Srivastava S, Bisht A, and Narayan N. Safety and security in smart cities using artificial intelligence—A review. In: Cloud Computing, Data Science & Engineering-Confluence, 2017 7th International Conference on. IEEE; 2017. pp. 130–133.

[19]   Leitão P, Colombo AW, and Karnouskos S. Industrial automation based on cyber-physical systems technologies: Prototype implementations and challenges. Computers in Industry. 2016;81:11–25.

[20]   AlTawy R, and Youssef AM. Security tradeoffs in cyber physical systems: A case study survey on implantable medical devices. IEEE Access. 2016;4:959–979.

[21]   Chen D, and Chang G. A survey on security issues of M2M communications in cyber-physical systems. KSII Transactions on Internet and Information Systems (TIIS). 2012;6(1):24–45.

[22]   Yongfu L, Dihua S, Weining L, *et al*. A service-oriented architecture for the transportation cyber-physical systems. In: Control Conference (CCC), 2012 31st Chinese. IEEE; 2012. pp. 7674–7678.

[23]   Tariq MU, Grijalva S, and Wolf M. A service-oriented, cyber-physical reference model for smart grid. In: Khaitan SK, McCalley JD, Liu CC, editors. Cyber Physical Systems Approach to Smart Electric Power Grid. Berlin: Springer; 2015. pp. 25–42.

[24]   Lee EA. Cyber physical systems: Design challenges. In: Object-oriented Real-time Distributed Computing (ISORC), 2008 11th IEEE International Symposium on. IEEE; 2008. pp. 363–369.

[25]   Lee I, Sokolsky O, Chen S, *et al*. Challenges and research directions in medical cyber–physical systems. Proceedings of the IEEE. 2012;100(1):75–90.

[26]   Burmester M, Magkos E, and Chrissikopoulos V. T-ABAC: An attribute-based access control model for real-time availability in highly dynamic systems.

In: Computers and Communications (ISCC), 2013 IEEE Symposium on. IEEE; 2013. pp. 143–148.

[27]   Al-Turjman F, Ever E, and Zahmatkesh H. Green femtocells in the IoT era: Traffic modeling and challenges–An overview. IEEE Network. 2017;31(6):48–55.

[28]   Al-Turjman F, Radwan A, Mumtaz S, *et al.* Mobile traffic modelling for wireless multimedia sensor networks in IoT. Computer Communications. 2017;112:109–115.

[29]   Al-Turjman F. Information-centric framework for the Internet of Things (IoT): Traffic modeling & optimization. Future Generation Computer Systems. 2018;80:63–75.

[30]   Nakamoto S. Bitcoin: A peer-to-peer electronic cash system [Working Paper]; 2008.

[31]   Minoli D, and Occhiogrosso B. Blockchain mechanisms for IoT security. Internet of Things. 2018;1:1–13.

[32]   Buterin V. A next-generation smart contract and decentralized application platform [White paper]; Ethereum Foundation, 2014.

[33]   Zhang L, Zheng Y, and Kantoa R. A Review of homomorphic encryption and its applications. In: Proceedings of the 9th EAI International Conference on Mobile Multimedia Communications. ICST (Institute for Computer Sciences, Social-Informatics and Telecommunications Engineering); 2016. pp. 97–106.

[34]   Al-Turjman F, and Alturjman S. Context-sensitive access in industrial Internet of things (IIoT) healthcare applications. IEEE Transactions on Industrial Informatics. 2018;14(6):2736–2744.

[35]   Sanduleac M, Eremia M, Toma L, *et al.* Energy ecosystem in smart cities—Privacy and security solutions for citizen's engagement in a multi-stream environment. In: Smart Cities Conference (ISC2), 2016 IEEE International. IEEE; 2016. pp. 1–4.

[36]   Weber RH. Internet of Things–New security and privacy challenges. Computer Law & Security Review. 2010;26(1):23–30.

[37]   Long YH, Tang ZH, and Liu X. Attribute mapping for cross-domain access control. In: Computer and Information Application (ICCIA), 2010 International Conference on. IEEE; 2010. pp. 343–347.

[38]   Bertino E, and Ghinita G. Towards mechanisms for detection and prevention of data exfiltration by insiders: Keynote talk paper. In: Proceedings of the 6th ACM Symposium on Information, Computer and Communications Security. ACM; 2011. pp. 10–19.

[39]   The 2017 U.S. State of Cybercrime Survey. Forcepoint and CSO and U.S. Secret Service and CERT Division of Software Engineering Institute at Carnegie Mellon University; 2018 [updated 2018 Jan 17; cited 2018 Apr 11]. Available from: https://insights.sei.cmu.edu/insider-threat/2018/01/2017-us-state-of-cybercrime-highlights.html.

[40]   Baracaldo N, and Joshi J. A trust-and-risk aware RBAC framework: Tackling insider threat. In: Proceedings of the 17th ACM symposium on Access Control Models and Technologies. ACM; 2012. pp. 167–176.

[41] Sicari S, Rizzardi A, Grieco LA, *et al*. Security, privacy and trust in Internet of Things: The road ahead. Computer networks. 2015;76:146–164.

[42] Li W, Song H, and Zeng F. Policy-based secure and trustworthy sensing for Internet of things in smart cities. IEEE Internet of Things Journal. 2017; 5(2):716–723.

[43] Chen R, Guo J, and Bao F. Trust management for SOA-based IoT and its application to service composition. IEEE Transactions on Services Computing. 2016;9(3):482–495.

[44] Wang JP, Bin S, Yu Y, *et al*. Distributed trust management mechanism for the Internet of things. In: Applied Mechanics and Materials. vol. 347. Trans Tech Publ. 2013; 2463–2467. Available from: https://www.scientific.net/ AMM.347-350.2463.bib

[45] Lize G, Jingpei W, and Bin S. Trust management mechanism for Internet of Things. China Communications. 2014;11(2):148–156.

[46] Samarati P, and de Vimercati SC. Access control: Policies, models, and mechanisms. In: International School on Foundations of Security Analysis and Design. Berlin, Heidelberg: Springer; 2000. pp. 137–196.

[47] Lampson BW. Protection. ACM SIGOPS Operating Systems Review. 1974;8(1):18–24.

[48] Graham GS, and Denning PJ. Protection: Principles and practice. In: Proceedings of the May 16–18, 1972, Spring Joint Computer Conference. ACM; 1972. pp. 417–429.

[49] Harrison MA, Ruzzo WL, and Ullman JD. Protection in operating systems. Communications of the ACM. 1976;19(8):461–471.

[50] Bell DE, and LaPadula LJ. Secure computer systems: Mathematical foundations. MITRE CORP BEDFORD MA; 1973.

[51] Biba KJ. Integrity considerations for secure computer systems. MITRE CORP BEDFORD MA; 1977.

[52] Jin X, Krishnan R, and Sandhu R. A unified attribute-based access control model covering DAC, MAC and RBAC. In: IFIP Annual Conference on Data and Applications Security and Privacy. Berlin, Heidelberg: Springer; 2012. pp. 41–55.

[53] Shafiq B, Joshi JB, Bertino E, *et al*. Secure interoperation in a multidomain environment employing RBAC policies. IEEE transactions on knowledge and data engineering. 2005;17(11):1557–1577.

[54] Pustchi N, and Sandhu R. MT-ABAC: A multi-tenant attribute-based access control model with tenant trust. In: International Conference on Network and System Security. Cham: Springer; 2015. pp. 206–220.

[55] Hu VC, Ferraiolo D, Kuhn R, *et al*. Guide to attribute-based access control (ABAC) definition and considerations (draft). Gaithersburg (MD): National Institute of Standards and Technology; 2013. NIST Special Publication 800-162.

[56] Bauer M, Boussard M, Bui N, *et al*. IoT reference architecture. In: Bassi A, Martin B, Martin F, *et al*., editors. Enabling Things to Talk. Berlin: Springer; 2013. pp. 163–211.

[57] Hernández-Ramos JL, Jara AJ, Marín L, *et al.* DCapBAC: Embedding authorization logic into smart things through ECC optimizations. International Journal of Computer Mathematics. 2016;93(2):345–366.

[58] Alshehri A, and Sandhu R. Access control models for cloud-enabled internet of things: A proposed architecture and research agenda. In: Collaboration and Internet Computing (CIC), 2016 IEEE 2nd International Conference on. IEEE; 2016. pp. 530–538.

[59] Gupta M, and Sandhu R. Authorization framework for secure cloud-assisted connected cars and vehicular internet of things. In: Proceedings of the 23nd ACM on Symposium on Access Control Models and Technologies. ACM; 2018. pp. 193–204.

[60] Bhatt S, Patwa F, and Sandhu R. Access control model for AWS Internet of Things. In: International Conference on Network and System Security. Cham: Springer; 2017. pp. 721–736.

[61] Akhuseyinoglu NB, and Joshi J. A risk-aware access control framework for cyber-physical systems. In: Collaboration and Internet Computing (CIC), 2017 IEEE 3rd International Conference on. IEEE; 2017. pp. 349–358.

[62] Venkatasubramanian KK, Nabar S, Gupta SK, *et al.* Cyber physical security solutions for pervasive health monitoring systems. In: User-Driven Healthcare: Concepts, Methodologies, Tools, and Applications. IGI Global; 2013. pp. 447–465.

[63] Venkatasubramanian KK, Mukherjee T, and Gupta SK. CAAC—An adaptive and proactive access control approach for emergencies in smart infrastructures. ACM Transactions on Autonomous and Adaptive Systems (TAAS). 2014;8(4):20.

[64] Gupta SK, Mukherjee T, and Venkatasubramanian K. Criticality aware access control model for pervasive applications. In: Pervasive Computing and Communications, 2006. PerCom 2006. Fourth Annual IEEE International Conference on. IEEE; 2006. pp. 257–261.

[65] Chen M, Chowdhury RA, Ramachandran V, *et al.* Priority queues and Dijkstra's algorithm. Computer Science Department, University of Texas at Austin; 2007.

[66] Joshi J, Ghafoor A, Aref WG, *et al.* Digital government security infrastructure design challenges. Computer. 2001;34(2):66–72.

[67] Baliosian J, and Cavalli A. An abstraction for the interoperability analysis of security policies. In: International Conference on Network and System Security. Cham: Springer; 2015. pp. 418–427.

[68] Joshi JB, Shyu ML, Aref W, *et al.* A multimedia-based threat management and information security framework. In: Web and Information Security. IGI Global; 2006. pp. 215–241.

[69] Gong L, and Qian X. Computational issues in secure interoperation. IEEE Transactions on Software Engineering. 1996;22(1):43–52.

[70] Lin D, Rao P, Bertino E, *et al.* An approach to evaluate policy similarity. In: Proceedings of the 12th ACM symposium on Access control models and technologies. ACM; 2007. pp. 1–10.

[71] Cho E, Ghinita G, and Bertino E. Privacy-preserving similarity measurement for access control policies. In: Proceedings of the 6th ACM workshop on Digital identity management. ACM; 2010. pp. 3–12.

[72] Rao P, Lin D, Bertino E, *et al.* Fine-grained integration of access control policies. Computers & Security. 2011;30(2-3):91–107.

[73] Mazzoleni P, Bertino E, Crispo B, *et al.* XACML Policy Integration Algorithms: Not to be confused with XACML Policy Combination Algorithms! In: Proceedings of the Eleventh ACM Symposium on Access Control Models and Technologies. ACM; 2006. pp. 219–227.

[74] Baracaldo N, Masoumzadeh A, and Joshi J. A secure, constraint-aware role-based access control interoperation framework. In: Network and System Security (NSS), 2011 5th International Conference on. IEEE; 2011. pp. 200–207.

[75] Cruz IF, Gjomemo R, and Jarzab G. An interoperation framework for secure collaboration among organizations. In: Proceedings of the 3rd ACM SIGSPATIAL International Workshop on Security and Privacy in GIS and LBS. ACM; 2010. pp. 4–11.

[76] Gouglidis A, Mavridis I, and Hu VC. Security policy verification for multi-domains in cloud systems. International Journal of Information Security. 2014;13(2):97–111.

[77] El Maarabani M, Hwang I, and Cavalli A. A formal approach for interoperability testing of security rules. In: Signal-Image Technology and Internet-Based Systems (SITIS), 2010 Sixth International Conference on. IEEE; 2010. pp. 277–284.

[78] Jung Y, and Joshi JB. CPBAC: Property-based access control model for secure cooperation in online social networks. Computers & Security. 2014;41:19–39.

[79] Bernabe JB, Ramos JLH, and Gomez AFS. TACIoT: Multidimensional trust-aware access control system for the Internet of Things. Soft Computing. 2016;20(5):1763–1779.

[80] Tang B, and Sandhu R. Cross-tenant trust models in cloud computing. In: Information Reuse and Integration (IRI), 2013 IEEE 14th International Conference on. IEEE; 2013. pp. 129–136.

[81] Tang B, and Sandhu R. Extending OpenStack access control with domain trust. In: International Conference on Network and System Security. Cham: Springer; 2014. pp. 54–69.

[82] Tang B, Sandhu R, and Li Q. Multi-tenancy authorization models for collaborative cloud services. Concurrency and Computation: Practice and Experience. 2015;27(11):2851–2868.

[83] Ferraiolo DF, Sandhu R, Gavrila S, *et al.* Proposed NIST standard for role-based access control. ACM Transactions on Information and System Security (TISSEC). 2001;4(3):224–274.

[84] Nasr PM, and Yazdian-Varjani A. Toward operator access management in SCADA system: Deontological threat mitigation. IEEE Transactions on Industrial Informatics. 2018;14(8):3314–3324.

[85] Bijon KZ, Krishnan R, and Sandhu R. A framework for risk-aware role based access control. In: Communications and Network Security (CNS), 2013 IEEE Conference on. IEEE; 2013. pp. 462–469.

Chapter 2
Impact of Internet of Things in smart cities
Bhawana Rudra[1]

A smart city uses various technologies to make the lives more easy and simple fulfilling the demands of the increasing population. Internet of Things (IoT) plays a major role in making the city smart. It takes the required input and helps to make the things associated with it smart. Some of the smart operations include water and energy management which is becoming scarce. The system can be deployed in the cities to make things work and save future resources. A smart city can be empowered to increase the quality of life of the people and improve the environment to sustain for a long time. Implementing a smart city with IoT and connected technology helps enhance the quality, performance and interactivity of urban services, optimize resources and reduce costs. The chapter briefly discusses what is the role of IoT in smart cities describing the basics of what is IoT and what comprises a smart city followed by smart city segments. Benefits of IoT and their impact on the smart city along with the national and international case studies. At the end of the chapter, some of the challenges associated with the IoT with respect to smart cities.

2.1 Introduction

It is estimated that there will be an increase of 70% in the urban population by 2050 based on the observations done by information and communications technology (ICT) [1–3]. It was observed in 1950, there was a net increase of 1.3 billion people in the small cities and expected to be double in the number in the medium (632 million) and large (570 million) cities by 2020 [3].

People are moving to urban areas in search of a better standard of living and job opportunities. This migration is leading to major issues like increased demand in natural resources, traffic congestion, water, sanitation and health care.

ICTs have come forward to provide a solution that is environment-friendly and economically feasible to solve some of the problems faced by the cities called smart cities. Some of them where they have suggested the solutions include waste management, water, transport infrastructure, energy, air quality testing and so on [4].

[1]National Institute of Technology Karnataka, Surathkal, Mangalore, India

2.2   Smart city

Smart cities are emerging with the emerging technologies of the smart solution in various sectors of society. This is trying to make the life of the citizens more easy and simple allowing the research and business to innovate. The main goals of the smart city are to provide quality life to the people along with the improvement in business and competition with various challenges, and the sustainability of the entire system has to be maintained. It connects physical, social and information communication infrastructure and helps to collect the data about the city [2,5]. A city is treated as intelligent, if it assists things like mobile communications, Internet and cloud computing to achieve fully aware, pervasive computing and converged applications. It adopts four characters like [6]:

1. Full of things include real-time sensing of things and connect to a network.
2. Full integration where the things are fully connected with full integration to work and provide the information and intelligent infrastructure and living environment.
3. Encouraging the individuals, business for the development of various applications for the city safety and steady stream of power.
4. Work together to collaborate with the critical systems present all over the city and coordinate to achieve the goal of society.

IoT plays a major role in deploying several services and making a city smart all over the world. The concept of IoT provides opportunities to manage devices and take an action based on the information gathered from various real-time devices. Efficient management along with the IoT devices will allow managing not only the devices but it allows to manage systems and the people and this will help for the development of the smart city. IoT has to reduce its cost and risk in developing its services. It will able to connect multiple heterogeneous systems all over the city. It should consider less time for the services to be deployed as a part of the initiative. The services offered by IoT should be secure and scalable providing new opportunities, able to create better service from smart connected data and devices [7]. IoT devices appear today to allow us not only to anticipate rather than simply react. Connected devices have been deployed in many areas like health, surveillance and transport. It has been estimated by the Chinese market that there will be an increase of 13.3% over the next five years. Many companies like Lenovo and Huawei are making heavy investment in the IoT sector. Many IoT systems like air monitoring or the smart home are developed for a particular task or they are closed but making the complete home or environment smart [8–10].

2.3   IoT technologies for smart cities

IoT is increasing day by day due to its promising technology and helping in solving many societal challenges. International Data Corporation (IDC) worldwide Internet Forecast has estimated that the usage of IoT will reach 30 billion by 2020 [11]. For the processing and storage of information in IoT, adopted the cloud platforms by the companies and have started developing the smart sensors with less cost has raised the IoT growth. Many large enterprises like Telefornica, Elliott management corporation and many other groups like Air Liquide, SUEZ, and so on have come forward and started investing for the research and development of IoT [12].

IoT deployment needs some common standards to operate the devices. The International Telecommunication Union (ITU), Institute of Electrical and Electronics Engineering (IEEE), IETF, ITU-Y. 2060 specifies some conditions for the machine interface oriented devices towards application various standards like IEEE 802.15.4 for link layer, IPv6 over low power Wireless Personal networks (6 LowPAN) for network layer and IPv6 routing protocol (RPL) standard for low power networks, Constrained application protocol (CoAP) to provide (HTTP) GET for querying and change the status at application level [13,14]. CoAP which relies on UDP protocol can provide communication security using Datagram Transport layer security. The various Operating Systems (OS) used in IoT is Contiki OS, Tiny OS, Windows10 IoT Core, Lite OS, etc. At the application layer of IoT, many applications have been developed; some among many are Nimbits, exosite One and Everything [15]. OASIS has issued many networks and massaging technologies for real-time systems of IoT. There are thousands of funding agencies and members to contribute which include IT consortium, Bosh, Intel, IBM, and many others for developing the enabling technologies of the industrial Internet. It is reported that the 19 groups and teams are working in different areas where some providing test beds, and some other working on the security of the device communication and so on [12].

The things involved in IoT for communication are mobile phones, appliances, etc are working to achieve an objective. The availability of various platforms allow to integrate with new devices, they communicate with each other, provides interoperability and scalability. Low-power standards are suitable for communication among various devices. Some networks are based on the location and distance are Home Area network (HAN) that use ZigBee, Dash7, and Wi-Fi and the components are connected to each other in a home. Wide Area Networks (WAN) provides communication between the utilities that are distributed among customers and implementation requires Fiber cable or Wireless like LTE or 3G. Field Area Networks (FAN) used to connect the substations and the customers [11,16,17].

IoT performs two tasks namely sensing and processing of those data. Speakthing is an IoT platform used for analyzing the gathered real-time data in the cloud. iOBridge is another technology used to connect to the cloud and accessed by web interfaces, and the data collected using this can be used for other web services. Some of the IoT technologies that are in use are explained as follows:

1. Radio Frequency Identification (RFID): This plays a vital role in the field of IoT. It dedicates the automatic detection with a single digital identity to any of the things associated with digital information. It provides the services in tracking the objects, health care, parking lots and smart grids. Each tag acts as a sensor as this does not have data which are written but also captures the environmental information [15,18–20].

2. Near Field Communication (NFC): It is used for bidirectional short-distance communication like smartphones and its range is up to a centimeter. NFC as a wallet will allow the smartphone as a personal card like a bank card and ID card.

Due to its bidirectional nature, this can be used to share data between various devices. If we place NFC at one position at home and connect with the central controller, the status or the position of the objects can be changed by checking the location [11,17,20].

3. Low-rate Wireless Personal Area network (LoWPAN): It covers a distance up to 10–15 km. It is one amongst the short-range radio technology. The power consumption is low and the battery life is nearly about 10 years. It uses the lowest two-layer protocols of physical and medium access, besides the upper-layer protocols which include ZigBee and 6LoWPAN [19,21].

   • ZigBee: It is an IEEE 802.15.4 standard suitable for creating WPAN and its various applications are medical device collection, wireless light switches, traffic management system, etc. It works well in limited ranges supporting billions of devices. A mechanism was designed for the transmission of IPv6 packets. To use ZigBee for various applications, special equipment is usually required like a coordinator, end devices of ZigBee and routers [11,17,20,22].

   • 6LoWPAN: It is used to adopt IPv6 communication. The problems of compatibility and the constrained nodes of the 6LowPAN is solved by adopting the compression format of IPv6 [5,6,20].

4. Wireless Sensor Networks (WSNs): These can be applied for various applications. The various data are gathered about people and objects and their movements if combined with RFID. They consist of interfaces, memory to store the information and power supply along with multiple sensors. The data are collected in an analog format and converted to digital using analog to a digital device. Data are processed using memory and microcontrollers present in them according to the user requirements. The data are transferred by radio interface and all these pieces of equipment involved in WSN require a power supply to work accordingly [17,23].

5. Dash7: This is best suited for military applications as it can penetrate through walls than 2.4 GHz and works at 433 MHz and appealing for HANs. It is of kilometer distance range. Some of the applications where it can be used are material monitoring, warehouse optimization and smart meter developments [24].

6. 3G and Long-Term Evolution (LTE): This standard is being used for mobile phones and data terminals. LTE is made available all over the world including the Third World countries. This is designed for long ranges, and can be applied for WSNs that require to work in long ranges. The major disadvantage of this is cost and implementation of the devices have to be specified [11].

7. Addressing: The IoT in the smart environment requires interconnection among objects and humans. For this, the identification of devices is required to fetch the desired results. Reliability, scalability and strength are the main requirements for establishing unique addressing of IoT [11,25].

8. Middleware: Middleware plays a major role in connecting the things with the applications layer. It aggregates the communication abilities and the functionalities of the devices involved in the communication [11].

9. Platforms for smart cities: Network drive smart cities communication platforms were introduced with new features for the Mobile to Mobile (M2M) communication in smart cities. These are used to communicate with heterogeneous

access technologies and the applications associated along with the suppliers. They help to communicate IoT with the real world. One such platform is open MTC, and the other standards involved in communication are Bluetooth, Mesh Networking, low-rate WPAN, high-rate WPAN, coexistence, body area networks, visible light communications, a key management protocol, routing, peer-aware communication and so on [26,27].

2.4 Smart city: IoT applications

It is estimated that the smart city market will reach to hundreds of billion dollars by 2020 with annual spending of 16 billion nearly [28]. The market includes some of the key industries having a connection with service sectors like smart mobility governance [29]. A smart city is not yet considered complete by all cities due to political, technical and financial barriers but still is emerging in its own way [30–32].

The urban infrastructure in smart cities includes the smart infrastructure, smart building and properties, smart industrial environment, smart city services, smart energy management, smart water management, smart waste management, etc. A brief explanation of all these and the subcategories as shown in Figure 2.1 are explained in detail.

2.4.1 Smart infrastructure

1. **Smart lighting** This helps to save energy consumption in indoor and outdoor scenarios. The light will increase the efficiency by making the light glow less when no one is present and glow high when detects someone in the path. It reduces maintenance and saves power. To control the light intensity of the light units using different sensors like light sensor and motion sensors. This will help to lower energy usage in lighting systems. It helps in daylight harvesting and save energy by dimming out sectors with no occupancies. Street light consumes more power than the other environments. The motion sensors in the light units will detect the traffic situation, weather conditions and time of the day, and based on the presence of the people, operation of the lights will be controlled [5,33–35].

2. **Connected streets** The smart city is capable of data acquisition and information delivery between devices including about traffic, road works, etc. Smart streets will help to use efficient resources and enhance public transportation and the urban landscape with the help of display boards. It will help in monitoring the energy consumption and the salubrity of the environment [5,36].

3. **Smart parking** IoT technology is used to find vacant parking spots in the parking area. The parking information is gathered by the sensor deployed on the parking spot and delivered to the server. The information then is delivered to the drivers to take the decision using various platforms like mobile phones or advertisement boards. The embedded sensors transmit the data on the timing and duration of the space used by the vehicle using the local signal processing. This reduces the CO emissions, and also reduces the congestion. This can be integrated with urban

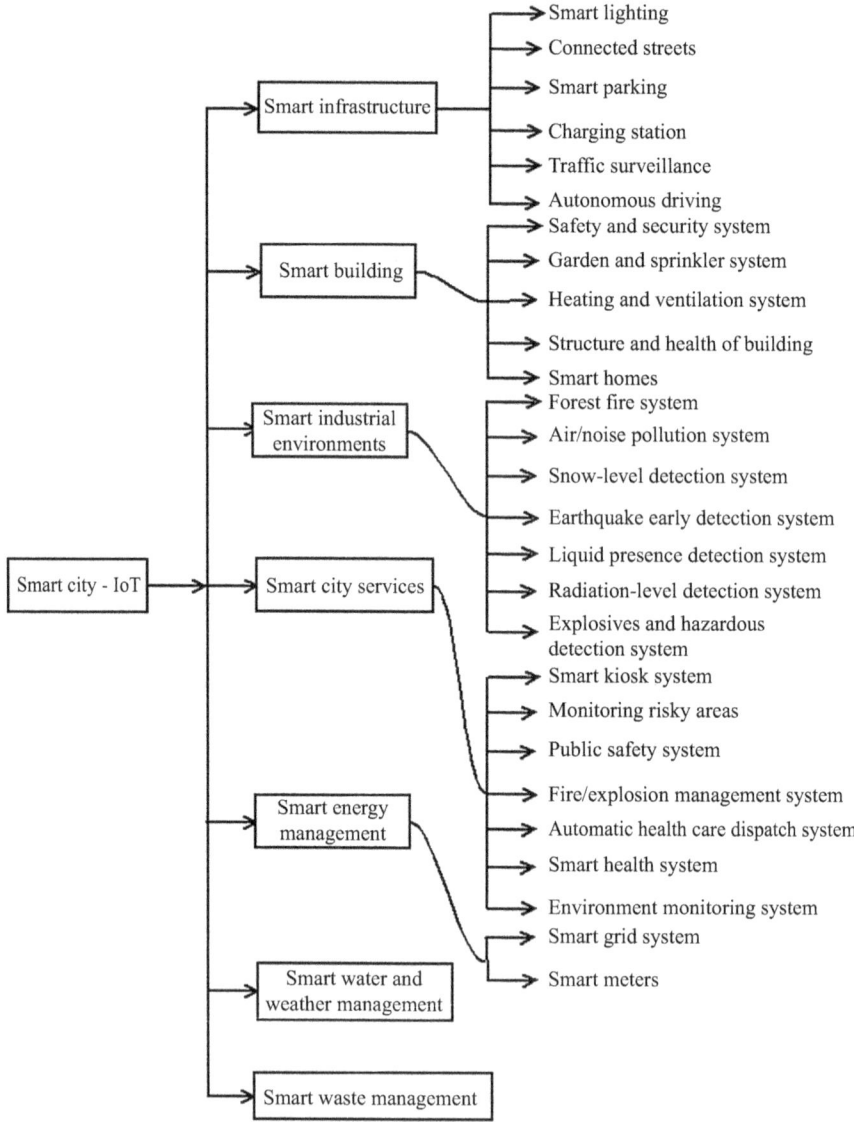

Figure 2.1   IoT applications

smart city infrastructure. It uses the technologies like RFID and NFC to realize the permits of the vehicles [6,12,36,37].

4.   **Connected charging stations** The charging stations must be made available at the parking slots, city fleets, building, malls, etc., in and around the city. This arrangement of the station will help the users to charge their electric vehicles on the go and can address the impact of the smart grid [12].

5. **Traffic surveillance and vehicular traffic** Surveillance will decrease the crime rate in the city; it detects the crime and gives the resolution with the help of camera images. The connected traffic signal will update about the traffic whereabouts thus reducing the congestion on the roads. These sensors can be connected to the central traffic management system, for the presentation of the complete traffic situation. Using the traffic pattern gathered along with the help of appropriate algorithm and the adjustments of the lights the system will generate challans against the rule violations. It also helps in reduction of fuel usage [12,22,33, 38–40].

6. **Autonomous driving** The introduction of autonomous driving will help the user to save time. This will help to speed up the traffic and save about 60% of parking space by allowing the cars to park close to each other. Many companies are introducing autonomous vehicles like Renault, Volvo and increasing the number of companies day by day. They are experimenting in real time by using various sensors that can trigger an alert when an accident or collisions or an unusual happens [12,41].

2.4.2 Smart building and properties

1. **Safety and security systems** The safety and security must be provided to the public and IoT will help to implement this by using biometrics, Surveillance camera alarms by not allowing the unauthorized parties to enter into the locality or homes. It can even detect and stops the perimeter access of the restricted properties [33,42].

2. **Smart garden and sprinkler systems** The sprinklers connected with IoT technologies and cloud can be used to water the plant based on the requirement. The devices can be used to know the level of fertilizers and moisture in the land. Based on the weather condition, sprinklers can schedule about the watering of the plant. Even the robot lawnmowers can be used to stop the overgrowing of the grass [12].

3. **Smart heating and ventilation system** The devices of this system can be used to monitor temperature, pressure, the humidity of the building, etc. A sensor network can be used to ensure appropriate heating and ventilation, can collect the information and optimize the systems that can improve efficiency and performance [33].

4. **Structural health of building** Continuous monitoring of the historical buildings is required to know the actual conditions of each building and identify the areas that are affected by the external agents. The introduction of IoT will provide building integrity measurements, by using sensors located in the various areas of the building. Some of the sensors that can be deployed for monitoring of the building are atmospheric sensors, building stress sensor, to measure the temperature, air pollution and the complete characterization of that building. To study the impact of earthquakes on the buildings, the combination of seismic and vibration sensors will help to achieve the target. The database where the concerned

information is stored can be made available to the public, and they will be aware of the care to be taken in preserving the historical buildings [33,43].

5. **Smart homes** The smart homes can be observed based on the data collected by the sensors. The demand response methods are applied for cautioning the users about the dangers if they cross the acceptable limit. IoT proves many applications for smart homes like smart TVs, security system, detection of fire, control the light and monitoring of the room temperature using various sensors. Some of the sensors are motion sensors, physical sensors, chemical sensors, leak/moisture detection, remote sensors, biosensors and so on [44]. The sensors will sense and send the data to the central controller present at home to take the best decision. The surveillance data can be used to predict future actions that may lose the comfort of standard living. Smart homes can be connected in the neighborhood by making the smart community [33,42].

2.4.3   Smart industrial environment

IoT plays a unique role in the industrial environment. Some of the applications of IoT are described as follows [44]:

1. **Forest fire detection** It monitors the combustion of gases and preemptive fire conditions. It is helpful in alert zones [6,12].
2. **Air/noise pollution** The technology can be used to monitor the quality of air in the areas where the people are in more number like crowded areas and parks. This will help them to be aware of the place and the situation in which they are living. Noise is seen in the form of acoustic pollution as CO in the air and IoT can be introduced to monitor the quality of air, noise at different places at different hours and help the municipality to introduce the measures [6,12,45].
3. **Snow-level monitoring & landslide and avalanche avoidance** It monitors the real-time condition of ski tracks that can prevent the avalanche situation [12]. The sensors monitor soil moisture, the density of the earth and vibrations if any along with dangerous patterns in the land conditions [12,46,47].
4. **Earthquake early detection** The deployment of sensors helps in detecting the tremors with the distributed control system at various places of tremors [12,46,47].
5. **Liquid presence** These sensors will help to detect the presence of liquid in any ground of the building, cloud data centers which will prevent the corrosion or any kind of breakdowns that may cause due to the presence of liquid [12].
6. **Radiation levels** These will detect the radiation levels in the nuclear power stations, and if found any leakage, alert will be sent [6].
7. **Explosive and hazardous gases** It detects the leakages in the gas levels in various factories like chemical factories, industrial environments and mines [45–47].

2.4.4   Smart city services

1. **Smart kiosk** The services that come under this are the 24/7 surveillance cameras and analysis, Wi-Fi services, digital signage for the announcements, free mobile

charging stations and the integration of environmental sensors. It even provides the information about the events nearby, retail stores, restaurants, etc. It will provide the map to the visitors for the additional data required in the mobile phones [48,49].

2. **Monitoring risky areas** The monitoring will be easy with the help of sensors in risky areas. It can send the alerts of any mishap in the nearby areas and send a message to avoid that particular area temporarily.
3. **Public safety** If installed in the areas of public organizations and the houses, it will send the information to fire station and the police when some mishap happens [33].
4. **Fire/explosion management** Based on the level of severity, fire sensors can automatically react to the conditions. It can send alert to the fire station, ambulance, blocking of the nearby streets from the traffic flow and coordinate within the rescue process with the help of drones and robots [6,12].
5. **Automatic health-care dispatch** With the help of technologies, medicines and drugs can be dispatched to the patients by providing 24/7 health care. These can also be used for calling the ambulance or patient pickup at the time of emergency [44,50].
6. **Smart health** Wireless Body Area Network (WBAN) can be used for monitoring the patients in the hospitals, work environments, etc. The device inserted into the body will communicate with the technologies available for communication like ZigBee, 6LowPAN and CoAP [50,51].
7. **Environment monitoring** The sensors that are placed around the city will monitor the electromagnetic field, temperature, toxic gases present in the air, any kind of combustion gases in the air which may detect fire and so on [52,53].

2.4.5   Smart energy management

1. **Smart grid** It is a digitally monitored system that delivers the gas or electricity to the required destinations from the source. This technology can be applied for residential and industrial areas. It uses intelligent and autonomous controllers for data management and maintains two-way communication. When applied for the power network, it yields a cost-effective power distribution and transmission [1,54–57].
2. **Smart meters** A smart meter measures the real-time information in the industrial or residential sectors. Energy consumption can be monitored, reports can be generated and even the dashboards can be accessed over the Internet [6,11,55,56].

2.4.6   Smart water and weather management

Conventional water methods to distribute the water are not suitable especially when the leakage is present in the pipeline and other related problems. Deployment of sensors at appropriate locations will monitor the damages and sends the information to the management. The quality and the quantity of the water can be known from lakes, reservoirs, aboveground pipelines and a storage tank with the help of sensors in the IoT environment. The water quality can be measured twice – once before storing it in the tank and next time after storing in the tank. With the help of weather sensors, rain cities can be recorded and can be analyzed for the development of local utilities, manage irrigation, decrease the excess usage of water, wastewater management. The localities can prepare to take care of controlling the raw sewage at the time of storms. The portable sensors can be deployed for monitoring of tap water, chemical leakage can be identified in rivers, and so on. The weather system uses various sensors to detect temperature, rain, solar radiation and speed of the wind to enhance the smart city efficiency [12,58,59].

2.4.7   Smart waste management

The municipality will get much help with the introduction of smart waste management system. The sensors attached to the garbage containers will help to detect the level of waste and send an alert to the municipality for the collection of the waste. This can improve the recycling of waste collected [12,60,61].

2.5   Applications: world wide

Not only internationally, but also in India, the effect of developing a smart city is in progress. The total urban population which got benefitted with the implementation of a smart city is 72 Mn and still, it is continuing. According to the reports in 2018, the total number of smart cities in India are 109 and it is still increasing day by day.

2.5.1   International-use cases

National Intelligence Council of the US has stated that the use of IoT is making the economic profits day by day [62]. In 2011, the interlinked objects are more than the humans in the environment. It was estimated that the objects interlink for communication has increased to 9 billion by 2012 and estimated to be 24 billion by 2020 [25]. It has provided that the IoT will be the major source for big data [63]. Many cities participated in the competition conducted by Intelligent Community Forum awards from 1999 to 2010 and the following cities were awarded and appreciated for their efforts in progressive ecosystem: Suwon and Seoul (South Korea), Taipei (Taiwan), Mitaka (Japan), Singapore, Waterloo and Calgary (Canada), Glasgow (Scotland), New York City and Georgia (USA) and Tehran (Iran) [64–69]. Some of the other countries that have come forward for the development of smart cities are as follows.

2.5.1.1   Amsterdam, The Netherlands

The project of smart city was launched in 2006. The connected lights played a major role in the dark period of time, able to invite for business and tourism. This also helped them to reduce the power consumption up to 80%, saving of about 130 billion euro by providing more visibility and safety to its citizens. The systems are interconnected via the Internet that led to more energy saving by detecting the failure automatically with the help of remote monitoring. The energy consumption is calculated by smart meters in recent years to know it accurately. To save power, the lights are dimmed during the low traffic hours and enhanced when required [70,71].

2.5.1.2   Chicago and New York, USA

The south side neighborhoods of Chicago started communicating with others using digital tools like web and mobiles to share information with the police to decrease the crime rate. The University of Chicago has created an awareness of spreading trust with the localities and other organizations. The information collected by the agents and police will be sent through a mobile App that is used to map with GPS for further action. In New York, to revitalize the city they used a 24/7 platform that will be available to the public for their citizen's safety. This is made available for any devices from anywhere at any time in the city. The platform integrates the information that is received from various locations and devices to build awareness among the people. Smart screens are made available at malls, stations to inform the citizens with the information that is relevant to their immediate proximity, local police, fire station, etc. [71,72].

2.5.1.3   Busan, South Korea

The government has provided job opportunities for the graduates and the economic growth using ICT. It has expanded the cloud infrastructure due to its good communication with the objects. The cloud connects the government, mobile application center that provides project and meeting rooms, consulting centers for start-up, an application for accessing municipal data from city geographic location and using a smart transport system. This has improved the life of its citizens by providing access to all the services. It allows the developers to develop apps through a shared platform by making the city to provide smart city services [71,73].

2.5.1.4   Nice, France

The city has implemented the Internet of energy (IoE), tested along with IP enabled technology studied the social and economic benefits to validate these. They observed that the data gathered by sensors can be used for smart parking and environmental monitoring using the traffic patterns. They provided four smart city services such as smart lighting, circulations, waste management and environment monitoring [71].

2.5.1.5   Padova City, Italy

The smart city of Padova is an example of public and private cooperation. The municipality has provided the required equipment and the budget to establish a smart city by allowing the University to implement the concept. Some of the implemented IoT devices for the collection of environment elements are CO level in the air, air temperature and humidity, noise, etc. The street light poles connected with sensors and Internet will help to collect data by means of wireless nodes. This concept has provided a gateway of implementing the IoT by considering the technological, social and political issues [6].

2.5.1.6   Sweden

Green IoT solution was incorporated with the help of smart sensing and cloud computing to have a more responsive and interactive city with the public and private sectors. The applications that are deployed with Green IoT are transportation, factory process optimization, environment monitoring, home security, etc. Uppsala city has installed IoT services like air pollution monitoring and traffic planning. Not only the service, but it has also taken care of data management, integration of green networking and techniques for low-power-efficient techniques [57].

2.5.1.7   Jiujiang City, China

The city has completed many projects like 6,000 points for the video surveillance system to capture video, smart transportation, emergency command system and smart urban management system, and many others were developed and deployed. A smart city is dependent on three factors and they are green, service-oriented and ubiquitous technologies. Although the city has Internet everywhere, the city is implanted with single object sensors connected to form things, for achieving the overall perception. With the IoT, it has increased the construction investments, and focus on providing facilities, resources, improving quality and service-level requirements. The construction of a fiber optic broadband network achieves the full coverage of urbanization areas. For the new buildings, residential quarters have been built to accelerate Fiber to the Premises (FTTH) as it can cover more than one million households. Construction of wireless city by providing multicity hotspots to strengthen the International and national communication system. Provide security management system, focus on energy saving and service innovation for achieving large scale deployment of devices [74].

Not only these countries but many others like Barcelona have also implemented the sensor technologies for evaluating the traffic flow that will help to design new bus networks and smart traffic management [64–68]. Stockholm has provided the whole city with fiber optic network [75]. Santa Cruz has used the network to analyze the information of criminal actions and predict the police requirement at that particular region [76]. Songdo, Korea, has constructed fully automated buildings, smart street lighting, and on [69]. Fujisawa in Japan has decreased the carbon footprint by 70% after implementation of the air monitoring system [17]. Norfolk, England, improved the delivery services, collection and system analysis for the municipality [17]. Kuala Lumpur city has 73 initiatives related to IoT with 24 strategic directions and eight Development thrusts are one of the developing city and on.

2.5.2   *National-use cases*

The increase in day-by-day economy of India states that there will be a rise of five times growth in urban centers. It is expecting a growth of 270 million workers and the job of 70%. Existing cities are extended which are almost like new towns. Some examples are Rohini, Dwarka and Narela as extensions to Delhi, Navi Mumbai to Mumbai, Salt Lake City to Kolkata, and Yelhanka and Kengeri to Bangalore, Noida, Greater Noida, Manesar, Pimpri-Chinchwad, Rajarhat, Dankuni, etc. This is raising many challenges in terms of quality living, jobs and so on. The adoption of smart cities is making the lives of people easy and simple. Many major cities of India which have initiated the smart city work as a part of development.

2.5.2.1   Mumbai City

The usage of CCTV network in Mumbai helped to catch the overstepping zebra crossing. This has decreased the chain snatching to 50% in 2016 and caught 14,000 traffic offenders. This also helped to generate 324 challans within 24 hours. Further, the Holi drunk and drive cases reduced to 31% in 2017. The city provides other smart services like Picture Intelligent unit, Vehicle tracking system, Collaborative monitoring, multiple command viewing centers, and video management analysis are some of them [75,77–79].

2.5.2.2   Jaipur City

The government has initiated the smart city mission to ensure 100% monitoring to create a safe city. For this, it has deployed 150 cameras that provide a 24/7 surveillance solution by making the city safe and smart. The city has increased the shared public transport by 45% which helped to reduce traffic jams in the city. The implementation of IoT in the city has helped to improve the mobility from 10% to 15%. For the citizens, it has opened 15 interactive kiosks and deployed 15 environmental sensors to protect the environment that gives the alarm if some unexpected happens. The city started managing the parking lot using the smart parking management solution. The city has opened two Remote Expert Government Services (REGS) to help the people remotely. It has opened 250 Wi-Fi access points allowing the citizens of Jaipur to access the Internet in an uninterrupted manner [75,77–79].

2.5.2.3   Nagpur City

Nagpur is the first city that has set up interactive kiosks at 100 locations all around the city. The city operation center provides various civic services and a central command control for security solutions. It has installed approximately 3,800 cameras for city surveillance at nearly 700 major junctions. It maintains a centralized infrastructure for Wi-Fi with 1,360 access points for 136 hot spot locations all over the city. Within a strip of 6 km, it provides smart lighting that saves energy, and smart transport maintaining the traffic by applying the smart traffic technique. Smart parking helps in saving fuel and time, and smart solid waste management and smart environmental sensors to sense the air and if any suspicious found immediately to warn the respective locality. The progress of smart city and its applications are still improving day by day [75,77,78,80,81].

2.5.2.4   Pune City

The government has initiated the smart city mission that providesWi-Fi to the city at various locations using some environmental sensors like Fire, Temperature weather, smoke etc.; Public announcement system for the safety of people; Emergency call box for providing emergency services; Variable message boards to display the messages like the nearest fuel bunk and distance of the road; and Smart city operating centre to decrease the crime rate and operate all the smart services provided for the welfare of the citizens [75,77,78,82].

2.5.2.5   Vizag City

The city has implemented the smart city concept by deploying some of the applications like surveillance system to decrease the crime rate, city governance system for smart light, smart waste management system, public announcement system, emergency call box, intelligent traffic system to control traffic in the city, smart water and energy managed by field teams, ambulance, call center citizen app for citizens for various purposes, etc. The implementation of the Drones for various services to provide monitor and create is in process, GVMC (ERP) service and citizen service centers are present to provide services to the citizens of Vizag. All these services and the field teams are maintained by city command and control center [75,77,78,83].

2.5.2.6   Hyderabad City

Hyderabad command control center provides the services like surveillance and collaborative monitoring, HAWK-Eye APP for public safety, portal and website services, traffic management systems, E-challan systems, body-worn camera solutions, various safety app, social media to help the need and SMS alerts, and smart city sensors to detect suspicious events in the city [75,77,78,84].

Not only these many other cities are emerging with the concept of a smart city. Many state governments are approaching with this concept to provide a comfortable and safe life for the people of the city.

2.6   Challenges and future directions

Implementation of IoT requires lots of planning and the economy. This is due to the various number of equipment involved in it. Each device is meant for a purpose and faces different challenges in implementing the desired task. The implementation of smart cities should place security on top priority. Some of the challenges include the following.

Security: The data gathering and evaluation of the system is performed on the platform of IoT. These systems are vulnerable to attacks like cross-site scripting. Its multi-tenancy leads to data leakage. the cities are adopting smart city concept which brings the security and trust of the citizens into consideration. Some of the security measures are to be implemented to gain the trust of citizens as well as to safeguard the critical infrastructure from cybersecurity attacks [58,85,86].

Heterogeneity: IoT devices are designed according to the specific application requirements. Each device performs a particular task. The authorities examine and define the needed hardware or software, and aggregation of these heterogeneous systems along with cooperating scheme is a challenging task [11,58].

The participation of a Large number of systems and support to all the technologies leads to reliability issue. Due to cars mobility, the interconnection of the system is not reliable [11,58].

Large scale: The IoT devices have to interact among a large number of distributed devices spread over in a large environment. The IoT platform allows to analyze and aggregate the information received from various devices. This data requires large storage and computational ability as it is received at higher rates which leads to storage issue, as well as the devices have delays related to dynamic connectivity [58,75,76].

Legal and social: IoT is a service and the service providers have to consider the local and international rules. If the applicants are allowed to choose and participate in the registration process, it will be more comfortable to develop the devices accordingly [87].

Big data: There are around 50,000,000,000 devices that communicate and transfer data to each other. The storage and analysis of this large amount of data has to be paid attention. In processing the IoT data, the major considerations to be noted are the number of devices, speed and variance [58].

Sensor networks: These can be considered as the technology for enabling IoT. They provide an understanding of the environment, measuring capabilities. Current development has provided chap devices for remote sensing. The best example we can consider is the mobile phones which consist of many sensors and this allows us to use these for various applications of IoT. But the challenge is the processing of large-scale information of the sensors with respect to energy and network constraints [88].

2.7 Conclusion

The growth of IoT in the current situation is observed to be increasing and its promising of more and smarter cities in the future. It not only provides the facilities in a more efficient manner but also make the quality of the citizen's life better. The implementation of IoT in all the urban areas will take time and effort but will be a profitable one. The governments are forward all over the globe to make the market friendly for the startups that are working on the IoT smart city solutions. Many smart city expos are being conducted to explore as well as to make aware of IoT usage all over the world. It will enhance the quality, performance, and interactivity of the services provided by the city thus reducing cost by utilization of resources in an optimized manner.

References

[1]   Harrison C, Eckman B, Hamilton R, *et al.* Foundations for smarter cities. IBM J Res Dev. 2010;54(4):350–365. Available from: http://dx.doi.org/10.1147/JRD.2010.2048257.

[2]   Behl A, Sheorey P, Nayak S, *et al.* Role of information and communications technology (ICT) in participatory democracy: A forerunner to an egalitarian society. In: Proceedings of the 10th International Conference on Theory and Practice of Electronic Governance. ICEGOV'17. New York, NY, USA: ACM; 2017. pp. 107–116. Available from: http://doi.acm.org/10.1145/3047273.3047375.

[3]   Meering C, and Balella HPEP. Smart Cities and the Internet of Things; 2016. Available from: https://www.hpe.com/h20195/v2/ GetPDF.aspx/4AA6-5129ENW.pdf (accessed on January 2016).

[4]   Smart City Examples. Available from: http://smartcitiescouncil.com/smart-cities-information-center/smart-city-examples.

[5]   Sikder AK, Acar A, Aksu H, *et al.* IoT-enabled smart lighting systems for smart cities. In: 2018 IEEE 8th Annual Computing and Communication Workshop and Conference (CCWC). 2018;639–645.

[6]   Zanella A, Bui N, Castellani A, *et al.* Internet of Things for smart cities. IEEE Internet of Things Journal. 2014;1(1):22–32.

[7]   Internet of Things in 2020: Roadmap for the Future; 2008. Available from: http://www.smart-systemsintegration.org/public/documents/publications/Internet-of-Things_in_ 2020_EC-EPoSS_Workshop_Report_2008_v3.pdf.

[8]   Guerrero-Ibáñez JA, Flores-Cortés C, and Zeadally S. In: Chilamkurti N, Zeadally S, Chaouchi H, editors. Vehicular Ad-hoc Networks (VANETs): Architecture, Protocols and Applications. London: Springer London; 2013. p. 49–70. Available from: https://doi.org/10.1007/978-1-4471-5164-7_5.

[9]   Hasan MZ, Al-Turjman F, and Al-Rizzo H. Optimized multi-constrained quality-of-service multipath routing approach for multimedia sensor networks. IEEE Sensors Journal. 2017;17(7):2298–2309.

[10]   Al-Turjman F, and Alturjman S. Confidential smart-sensing framework in the IoT era. The Journal of Supercomputing. 2018;74:5187–5198.

[11]   Talari S, Shafie-khah M, Siano P, *et al.* A review of smart cities based on the Internet of Things concept. Energies. 2017 03;10:1–23.

[12]   Hammi B. IoT technologies for smart cities. IET Networks. 2018;7:1–13(12). Available from: https://digital-library.theiet.org/content/journals/10.1049/iet-net.2017.0163.

[13]   Generation networks frameworks N fam. Recommendation ITU-T Y.2060, Overview of the Internet of things. International Telecommunication Union, Technical report, 2012.

[14]   Bello O, Zeadally S, and Badra M. Network layer inter-operation of device-to-device communication technologies in Internet of Things (IoT). Ad Hoc Netw. 2017;57(C):52–62. Available from: https://doi.org/10.1016/j.adhoc.2016.06.010.

[15]   Talari S, Shafie-khah M, Siano P, *et al.* A review of smart cities based on the Internet of Things concept. Energies. 2017;03(10):1–23.

[16]   Strategic Opportunity Analysis of the Global Smart City Market; Available from: http://www.egr.msu.edu/~aesc310-web/resources/SmartCities/Smart%20City%20Market%20Report%202.pdf.

[17]   Hancke GP, Silva BH, and Hancke GP Jr. The role of advanced sensing in smart cities. Sensors. 2012;1:393–425.

[18]   Zeadally S, Hunt R, Chen YS, *et al.* Vehicular ad hoc networks (VANETS): Status, results, and challenges. Telecommunication Systems. 2012;50(4): 217–241. Available from: https://doi.org/10.1007/s11235-010-9400-5.

[19] Kosmatos EA, Tselikas ND, and Boucouvalas AC. Integrating RFID and smart objects into a unified Internet of Things architecture. Adv Internet Things. 2011.

[20] Yaqoob I, Hashem IAT, Mehmood Y, *et al.* Enabling communication technologies for smart cities. IEEE Communications Magazine. 2017 January; 55(1):112–120.

[21] Niyato D, Hossain E, and Camorlinga S. Remote patient monitoring service using heterogeneous wireless access networks: Architecture and optimization. IEEE Journal on Selected Areas in Communications. 2009 May;27(4):412–423.

[22] Al-Turjman F. Information-centric framework for the Internet of Things (IoT): Traffic modeling & optimization. Future Generation Computer Systems. 2017;09;80.

[23] Alamri A, Ansari WS, Hassan MM, *et al.* A survey on sensor-cloud: Architecture, applications, and approaches. International Journal of Distributed Sensor Networks. 2013;9(2):917923. Available from: https://doi.org/10.1155/2013/917923.

[24] Medagliani P, Leguay J, Duda A, *et al.* Internet of Things Applications from research and innovation to market deployment. In Bringing IP to Low-Power Smart Objects: The Smart Parking Case in the CALIPSO Project. 2014; pp. 287–313.

[25] Gubbi J, Buyya R, Marusic S, *et al.* Internet of Things (IoT): A Vision, Architectural Elements, and Future Directions. CoRR. 2012;abs/1207.0203. Available from: http://arxiv.org/abs/1207.0203.

[26] Elmangoush A MTSWe and Alhazmi A. Towards unified smart city communication platforms. In Proceedings of the Workshop on Research in Information Systems and Technologies, 16 October 2015.

[27] 15 ISGP. : Wireless PANs. Available from: https://standards.ieee.org/about/get/802/802.15.html.

[28] Pike Research on Smart Cities. Available from: http://www.pikeresearch.com/research/smart-cities.

[29] M Dohler XV I Vilajosana, and LLosa J. Smart Cities: An Action Plan, Barcelona Smart Cities Congress 2011; Available from: Barcelona, Spain, Dec. 2011.

[30] Vilajosana I, Llosa J, Martínez B, *et al.* Bootstrapping smart cities through a self-sustainable model based on big data flows. IEEE Communications Magazine. 2013 June;51(6):128–134.

[31] Mulligan C, and Olsson M. Architectural implications of smart city business models: An evolutionary perspective. IEEE Communications Magazine. 2013 June;51(6):80–85.

[32] N Walravens PB. Platform business models for smart cities: From control and value to governance and public value. IEEE Communications Magazine. 2013;51(6):72–79.

[33]   Park E, del Pobil AP, and Kwon S. The role of Internet of Things (IoT) in smart cities: Technology roadmap-oriented approaches. Sustainability. 2005; 10:13–88.

[34]   Escolar S, Carretero J, Marinescu MC, *et al.* Estimating energy savings in smart street lighting by using an adaptive control system. International Journal of Distributed Sensor Networks. 2014;10(5):971–587. Available from: https://doi.org/10.1155/2014/971587.

[35]   Mllner R, and Riener A. An energy-efficient pedestrian-aware smart street lighting system. Int J Pervasive Computing and Communications. 2011;7: 147–161.

[36]   Kastner W, Neugschwandtner G, Soucek S, *et al.* Communication systems for building automation and control. Proceedings of the IEEE. 2005;93: 1178–1203.

[37]   Lu R, Lin X, Zhu H, *et al.* SPARK: A new VANET-based smart parking scheme for large parking lots; 2009. pp. 1413 – 1421.

[38]   Wang Jt, Chen Db, Chen Hy, *et al.* On pedestrian detection and tracking in infrared videos. Pattern Recogn Lett. 2012;33(6):775–785. Available from: http://dx.doi.org/10.1016/j.patrec.2011.12.011.

[39]   Damen D, and Hogg D. Detecting carried objects from sequences of walking pedestrians. IEEE Transactions on Pattern Analysis & Machine Intelligence. 2012;34(6):1056–1067.

[40]   Hasan MZ, and Al-Turjman F. Evaluation of a duty-cycled asynchronous X-MAC protocol for vehicular sensor networks. EURASIP Journal on Wireless Communications and Networking. 2017;2017(1):95. Available from: https://doi.org/10.1186/s13638-017-0882-7.

[41]   Renault NEXT TWO et la vie à bord hyper-connectée pour tous Renault TR; 2014.

[42]   Mano LY, Faial BS, Nakamura LHV, *et al.* Exploiting IoT technologies for enhancing health smart homes through patient identification and emotion recognition. Computer Communications. 2016;89-90:178–190.

[43]   Lynch J, and Loh K. A summary review of wireless sensors and sensor networks for structural health monitoring. The Shock and Vibration Digest. 2006;38: 91–128.

[44]   Al-Turjman FM. Intelligence in IoT-enabled Smart Cities. Imprint CRC Press. 2018; p. 242.

[45]   Maisonneuve N, Stevens M, Niessen M, *et al.* Citizen noise pollution monitoring; 2009. pp. 96–103.

[46]   Yick J, Mukherjee B, and Ghosal D. Wireless sensor network survey. Comput Netw. 2008;52(12):2292–2330. Available from: http://dx.doi.org/10.1016/j.comnet.2008.04.002.

[47]   Lazarescu M. In: Wireless Sensor Networks for the Internet of Things: Barriers and Synergies; 2017. pp. 155–186.

[48]   Rehg JM, Loughlin M, and Waters K. Vision for a smart kiosk. In: Proceedings of IEEE Computer Society Conference on Computer Vision and Pattern Recognition; 1997; pp. 690–696.

[49]   Nam T, and Pardo TA. Smart city as urban innovation: Focusing on manage-
         ment, policy, and context. In: Proceedings of the 5th International Conference
         on Theory and Practice of Electronic Governance. ICEGOV'11. New York,
         NY, USA: ACM; 2011. pp. 185–194. Available from: http://doi.acm.org/
         10.1145/2072069.2072100.
[50]   ICSIS. Available from: http://standards.ieee.org/getieee802/download/802.15.
         4-2011.pdf.
[51]   4e (TG4e) ICSIWTG. Available from: http://www.ieee802.org/15/pub/
         TG4e.html.
[52]   FP7-ENVIRONMENT program EcoWeb a dynamic e-dissemination platform
         for EU eco-innovation research results. Available from: http://ecoweb-
         project.info/.
[53]   A Castellani ARTF S Loreto, and Dijk E. Best Practices for HTTP-CoAP
         Mapping Implementation; 2013. Available from: draft-castellanicore-
         http-mapping-07.s.l.:IETF.
[54]   Siano P. Demand response and smart grids: A survey. Renewable and
         Sustainable Energy Reviews. 2014 02;30:461–478.
[55]   Zhabelova G, and Vyatkin V. Multiagent Smart grid automation Architecture
         based on IEC 61850/61499 intelligent logical nodes. IEEE Transactions on
         Industrial Electronics. 2012;59(5):2351–2362.
[56]   Yun M, and Yuxin B. Research on the architecture and key technology of
         Internet of Things (IoT) applied on smart grid. 2010 International Conference
         on Advances in Energy Engineering. 2010; pp. 69–72.
[57]   Ahlgren B, Hidell M, and Ngai ECH. Internet of Things for smart cities:
         Interoperability and open data. IEEE Internet Computing. 2016;20(6):52–56.
         QC 20170209.
[58]   Botta A, de Donato W, Persico V, *et al.* Integration of cloud computing and
         Internet of Things: A survey. Future Gener Comput Syst. 2016;56(C):684–700.
         Available from: https://doi.org/10.1016/j.future.2015.09.021.
[59]   Internet of Things TUI. Available from: http://datasmart.ash.harvard.edu/
         news/article/the-urbaninternet-of-things-727.
[60]   Fujdiak R, Masek P, Mlynek P, *et al.* Using genetic algorithm for advanced
         municipal waste collection in smart city. 2016 10th International Symposium
         on Communication Systems, Networks and Digital Signal Processing
         (CSNDSP). 2016; pp. 1–6.
[61]   Pala Z, and Inan N. Smart parking applications using RFID technology. 2007
         1st Annual RFID Eurasia. 2007; pp. 1–3.
[62]   National Intelligence Council. Disruptive Civil Technologies: Six Tech-
         nologies With Potential Impacts on U.S. Interests Out to 2025 (Apr. 2008).
         Available at https://fas.org/irp/nic/disruptive.pdf.
[63]   Petrolo R, Mitton N, Soldatos J, *et al.* Integrating wireless sensor networks
         within a city cloud. In: Eleventh Annual IEEE International Conference on
         Sensing, Communication, and Networking Workshops, SECON Workshops
         2014, Singapore, 30 June–3 July, 2014; 2014. pp. 24–27. Available from:
         https://doi.org/10.1109/SECONW.2014.6979700.

[64] de Premsa | El Web de la Ciutat de Barcelona SCS. Available from: http://ajuntament.barcelona.cat/premsa/tag/smart-city/ (accessed on 24 February 2017).

[65] Laursen ABwSCS L City Saves Money. Available from: https://www.technologyreview.com/s/532511/barcelona-smart-city-ecosystem/ (accessed on 24 February 2017).

[66] iTunes | apps4BCN | All the Apps You Need for Barcelona! TMBBA. Available from: http://apps4bcn.cat/en/app/tmbapp-metro-bus-barcelona/111 (accessed on 24 February 2017).

[67] for Barcelona | Apps4bcn | All the Apps You Need for Barcelona! TTITBAS. Available from: http://apps4bcn.cat/en/apps/index/Category:transport-i-tr-nsit (accessed on 24 February 2017).

[68] the Apps You Need for Barcelona! USB. Available from: http://apps4bcn.cat/en/app/urbanstep-barcelona/110 (accessed on 24 February 2017).

[69] kit IRT. Available from: http://www.ictregulationtoolkit.org/practice_note?practice_note_id=3244 (accessed on 24 February 2017).

[70] ASC. Available from: https://amsterdamsmartcity.com/ (accessed on 24 February 2017).

[71] The Internet of Everything for Cities TI. Available from: http://pie.pascalobservatory.org/sites/default/les/ioe-smart-city_pov.pdf

[72] City of Chicago Digital Roadmap to Improve Quality of Life. Available from: http://www.cityofchicago.org/city/en/depts/mayor/press_room/press_releases/2013/september_2013/mayor_emanuel_releasescityofchicagosfirstevertechnologyplan.html, Apr 2015.

[73] Strickland, E. Cisco bets on South Korean smart city, IEEE Spectrum, 2011;48:11–12.

[74] Liu Z, Ying W, Qin X, and Tao Y. Study on smart city construction of jiujiang based on IOT technology; IOP Conference Series: Earth and Environmental Science. 2017;69. doi:10.1088/1755-1315/69/1/012105.

[75] kit IRT. Available from: http://www.ictregulationtoolkit.org/practice_note?practice_note_id=3244.

[76] Writer SBSSMGiFSMoSCPPP. Available from: http://www.santacruzsentinel.com/article/zz/20120226/NEWS/120227300 (accessed on 24 February 2017).

[77] Smart Cities in India. Available from: https://www.rvo.nl/sites/default/files/Smart%20Cities%20India.pdf.

[78] Hoelscher K. The evolution of the smart cities agenda in India. International Area Studies Review. 2016;19(1):28–44. Available from: https://doi.org/10.1177/2233865916632089.

[79] Preksha Pandey and Akshay Kumar Pandey. vol. 23; (February. 2018). pp. 57–65. Available from: http://apps4bcn.cat/en/apps/index/Category:transport-i-tr-nsit (accessed on 24 February 2017).

[80] India's Heart Nagpur. Available from: https://www.unescap.org/sites/default/files/Nagpur%20Smart%20City.pdf (accessed on 24 February 2017).

[81] Kandpal V. A case study on smart city projects in India: An analysis of Nagpur, Allahabad and Dehradun. In: Companion Proceedings of The Web Conference

2018. WWW'18. Republic and Canton of Geneva, Switzerland: International World Wide Web Conferences Steering Committee; 2018. pp. 935–941. Available from: https://doi.org/10.1145/3184558.3191522.

[82]   Corporation PSCD. Available from: https://punesmartcity.in/wp-content/uploads/2018/10/Annual-report-2017-18.pdf (accessed on April 2017).

[83]   City VS. Available from: http://smartcities.gov.in/upload/uploadfiles/files/VishakapatnamSCP.pdf (accessed on April 2018).

[84]   Leveraging the Power of ICT for Good Urban Management. Available from: http://icrier.org/pdf/V_Srinivas_Chary_Smart.pdf.

[85]   The Internet of Things: Privacy I, Security in a Connected World Technical report US FederalTrade Commission (FTC).

[86]   Jing Q, Vasilakos AV, Wan J, *et al.* Security of the Internet of Things: Perspectives and challenges. Wirel Netw. 2014;20(8):2481–2501. Available from: http://dx.doi.org/10.1007/s11276-014-0761-7.

[87]   Atkins C, Koyanagi K, Tsuchiya T, *et al.* A cloud service for end-user participation concerning the Internet of Things. In: 2013 International Conference on Signal-Image Technology & Internet-Based Systems (SITIS). Los Alamitos, CA, USA: IEEE Computer Society; 2013. pp. 273–278. Available from: https://doi.ieeecomputersociety.org/10.1109/SITIS.2013.53.

[88]   Zhao F. Sensors meet the cloud: Planetary-scale distributed sensing and decision making. In: Proceedings of the 9th IEEE International Conference on Cognitive Informatics, ICCI 2010, 7–9 July 2010, Beijing, China; 2010. p. 998. Available from: https://doi.org/10.1109/COGINF.2010.5599715.

Chapter 3

IoT-based smart water

Hitesh Mohapatra[1] and Amiya Kumar Rath[1,2]

This chapter primarily focuses on the role of the Internet of Things (IoT) in the smart water (SW) conversion process. The several problems like leakage detection, efficient water distribution, remote water monitoring, etc., can be addressed by using IoT with the combination of information and communication technology (ICT). In this line of thought, we propose a smart water system (SWS) layer architecture to ensure the proper utilization of natural or man-made resources. The hygiene water is the birthright of every human being and to ensure it for a future generation the SWS is a proven model.

3.1 Smart city

The IoT is a new emerging trend which emerged with information technology for day-to-day use. The IoT concept was first proposed by Kevin Ashton in 1991; since that time the IoT is one of the most relevant technologies in the many applications. IoT is a global network infrastructure which consists of *things/objects* where these things have virtual personalities, identities, and attributes [1,2]. The properties like self-configuring, self-healing, and interoperable communication bring wide acceptability from various sectors of society [3]. According to a survey by Gartner [4], by 2020 there will be 26 billion devices which will be on the IoT network. The IoT technology is enabled with various advanced features such as intelligent monitoring, positioning [5], tracking based on identification, etc., and it extends its involvement in the various sectors of society like smart city and its application. The various applications of the smart city such as SW, smart grid [6], and smart agriculture, these all are impossible without IoT and ICT [7].

The discussion begins with a question, i.e. why does normal water need to convert into SW? The answer to this question may be available by asking another cross-question, i.e. why people are moving for urbanization? The very fundamental instinct of every living being is to *access* [8]. As far as human are concerned, being an intelligent species on earth human wants to access many things such as

---

[1]Computer Science & Engineering, Veer Surendra Sai University of Technology, Burla, Odisha, India
[2]Current affiliation: Deputed to National Assessment and Accreditation Council (NAAC), Bangalore, India

job; earning livelihood; necessities of life like house, food, water [9], and electricity; infrastructures like school, healthcare service, market place, sanitary system, and garbage disposal system; access to education, knowledge, information, and technical advances; access to people, to social, culture, religion, and communities; access to a place where humans have rights and code of conduct which is again regulated by an institution like court or government; and a place which provides a certain degree of security, predictability, stability, and sustainability from disasters either natural or man-made [10]. Cities are the places which provide all the above demands of mankind with a certain degree. It provides opportunities; worldwide connections either physical or logical via transport services (e.g. aeroplane, train, and bus), Internet, mobile phones, etc.; entertainment through theatre, cinema halls, etc.; and the desire to access all these facilities compelling people to move towards urban. Currently, 50% of the total population is living in urban place whereas it is expected to grow up to 66% by 2050 [11].

The first question is, what is meant by a *smart city*? The answer to this question is, there is no such universally accepted common definition exists of a smart city. It means different things to different people. Therefore, the conceptualization of smart city varies from city to city and country to country, depending on the level of development; willingness to change and reform; and resources and aspirations of the city residents. A smart city would have a different connotation in India than, say, Europe. Even in India, there is no common way of defining a smart city. Figure 3.1 illustrates few important aspects of smart cities.

Some definitive boundaries are required to guide cities in this mission. In the imagination of any city dweller in India, the picture of a smart city contains a wish list of infrastructures and services that describes a citizen's level of aspiration. According to the aspirations and needs of the citizens, urban planners ideally aim at developing

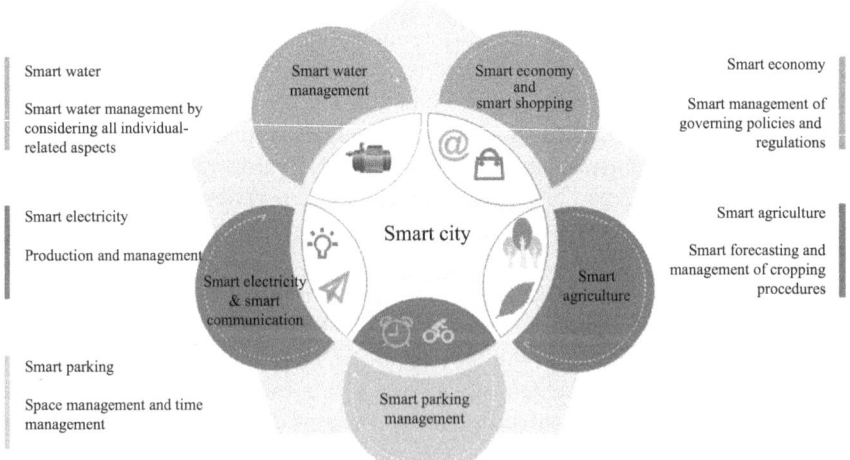

Figure 3.1   Smart city applications

the entire urban ecosystem, which is represented by the four pillars of comprehensive development [12]. The four pillars are institutional, physical, social, and economic infrastructure. This can be a long-term goal and cities can work towards developing such comprehensive infrastructure incrementally, adding on layers of smartness.

In the approach of the Smart Cities Mission (SCM), the objective is to promote cities that provide core infrastructure and give a decent quality of life to its citizens, a clean and sustainable environment, and application of Smart Solutions. The focus is on sustainable and inclusive development, and the idea is to look at compact areas and create a replicable model which will act as a lighthouse to other aspiring cities. The SCM of the government is a bold, new initiative. It is meant to set examples that can be replicated both within and outside the smart city, catalysing the creation of similar smart cities in various regions and parts of the country.

The core infrastructure elements in a smart city would include [13] the following:

1. adequate water supply;
2. assured electricity supply;
3. sanitation, including solid waste management;
4. efficient urban mobility and public transport;
5. affordable housing, especially for the poor;
6. robust IT connectivity and digitalization;
7. good governance, especially e-Governance and citizen participation;
8. sustainable environment;
9. safety and security of citizens;
10. health and education.

As far as smart solutions are concerned, an illustrative list is given next. This is not, however, an exhaustive list, and cities are free to add more applications.

3.1.1 IoT in traditional city

To accommodate these new demands from cities and urbanites, the IoT can be used as a tool which will reduce the cost, improve the interaction, and enhance the service. The role of IoT can be extended from the domestic environment to the harsh and hostile zone. For example, IoT can be used as a tool to avoid traffic congestion. The gradual increase in urbanite count leads to a serious traffic problem in the cities. Fortunately, the availability of IoT brings smart solutions like smart traffic signalling which accommodates the various density of traffic as per time and day concern. The other side of traffic management is reliable transportation which also can be taken care of by the use of IoT. For example, deployment of cameras and other connected devices at various locations like within the bus and at the bus stop will be helpful to provide a secure and reliable transport service to the citizens. There is another major dimension of IoT is smart buildings where the energy (electricity/water) can be utilized in a proper manner. The study says that the traditional building is wasting 30% of their total energy because of inadequate management. The smart building can keep track of several factors such as lighting, cooling, heating, and safety which will ultimately save enormous amount of energies and will provide a better comfortable lifestyle to the consumers. In urban place, there is another sector which also demands significant attention, i.e. public safety. From several existing ways, the two most popular method is the installation of camera and smart lighting. The deployment of the camera with connected devices at various locations like traffic point, streets, apartments, and shopping malls produces thousands of hours of video footage. When these videos are analyzed by video analytical software, it produces profound proofs for many important events. The smart lighting is one of the significant technical advancement which helps to conserve electricity. These smart LED lights embedded with motion sensor, brightness sensor, small communication network, and LEDs. The motion sensors keep tracks of passengers, vehicles, etc. According to the motion of the objects, the light gets on and off [14]. Like this many instances available in existing literature where the problems with traditional cities can be solved by the help of IoT and ICT.

3.1.2 Myths about smart cities

One thing which is very clear is that the cities have very complex architecture and to bring a specific change in it is extremely difficult. The imagination about a change is not a difficult task but in practice how the people will respond to those changes is again more complex. There are, for example, many counter questions always in ready state to attack the changes:

- Why people will motivate to adopt the change?
- How this option is better than the rest of proposed ideas?
- Is that good or bad?
- What are the probable side effects of this new idea?
- How long these ideas will sustain?
- What kind of benefits these ideas will give?
- Either these changes are specific for one sector or for all?
- What kind of side effects these changes will invite?

In case of real innovation, it is not possible to answer all these questions but at the same time these questions must and should be considered during planning and designing phase. There is another factor called misconception/myths which creates many additional challenges. In this section, we will discuss few such misconceptions or myths about the smart cities.

Myth 1: Everybody wants smart city: If we consider the total population of earth, then till date only 50% of total population have access to the Internet where 2 billion from developing countries and 89 million from least developed countries [15]. The statistical presentation of the Internet users is illustrated in Figure 3.2. The population without internet access are mostly busy with their own social life, family, and job. This point matters because what we are discussing in the World Smart City Forum never matters to this rest 50% of population. That's why Simon Giles of Accenture rightly quoted that smart cities industry not doing good enough in selling of ideas to the wide audience. This problem can be solved by spreading awareness among all types of population primarily in rural belts of

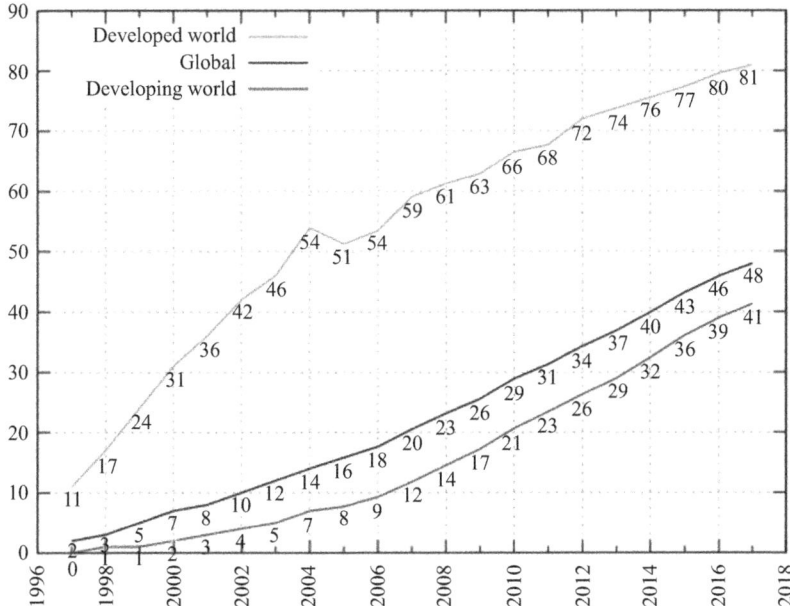

Figure 3.2 *Internet users per 100 inhabitants*
Source: https://upload.wikimedia.org/wikipedia/commons/2/29/
Internet_users_per_100_inhabitants_ITU.svg

the country. Because when this population will shift to urban place, they cannot use the smart services of smart cities which again will bring disturbances in law and order.

Myth 2: Assuming that applying technologies for smart cities is new: We must not forget the historical perspective which reminds about human evolution with technology and cities in single process. The presence of technology can be marked since the man-made tools from stone and woods. Since that time till today, the human race embarked with the process of socio-technological evolution. The only advancement what we have experiencing nowadays is digitalization of exiting methods.

Myth 3: Smart cities are by-product of inhuman technologies which brings risk: The major challenge behind every innovation is to bring the impacts and benefits of the innovation to an acceptable form by the people or communities. The innovation process need to be more focused on need and behaviours of the people rather than planning and designing [4]. The innovation of smart cities will be successful when the people will get the benefits of their co-creation.

Myth 4: Smart cities produces smart citizen: During a one-day workshop on SW, I asked one question, why it was that so many new urban developments seem not to take adequately into account the natural behaviour of the people expected to use them? The response of this question is human behaviours always adapts the best from its surrounding so, when the environment changes it takes time to make the people smart enough to accept the changes.

Myth 5: Business as usual will deliver the result: The total population is the combination of two categories of people that are rich and poor people. The demand of these two categories from smart cities drastically differs. That's why currently many business leaders, entrepreneur, innovators, and activists are thinking for new strategies to model their business which can reach both of these categories. The mission and vision of smart cities must be planned and designed by looking at both the aspects, i.e. poor and rich population demands in a balanced way; then only, the smart cities business will perform well in the future.

3.1.3 IoT for communication in smart cities

In this previous section, we had discussed the following: What is the smart city? Why it is important? What is the reality? But, in this particular subsection we need to understand either, smart city is just about deployment of modern gadgets or anything else also required to make this deployment meaningful? The answer to this question is this total deployment is meaningful when it will be supported by an efficient communication model. Selecting a solution without proper communication model will be costlier in the future. The failure in communication model can bring disaster. The very recent example is the strike of cyclone *FANI* at the coastal belt of India in 2019. The early alarm by India Meteorological Department (IMD) saved millions of life of various states like Odisha, India, and of coastal Bagerhat district, Bangladesh. In preparation for the storm's impact, the state government of Odisha evacuated over 1.2 million residents from vulnerable coastal areas and moved them to higher ground and into cyclone shelters built a few miles inland. The authorities deployed around a thousand emergency workers and 43,000 volunteers in these efforts. It sent out 2.6 million text messages to warn of the storm in addition to using television, sirens, and public-address systems to communicate the message. About 7,000 kitchens were operated to feed evacuees in 9,000 storm shelters [16]. Like this, there are many examples that exist in history of rescue which were possible only because of timely communication from remote end sensors. Table 3.1 represents the question and answer (Q&A) for proper designing and planning of IoT-based communication model.

3.2 Smart water

The water is a basic need for all mankind on a daily basis. The future water scarcity needs an efficient and immediate solution. In the twenty-first century, technological advancement is one of the key aspects of modern urban life. The revolutionized technical growth is now matured enough to provide a sustainable smart city with

Table 3.1   Q&A session for reliable communication

Interest	Questions	Answers
Performance	Is the system scalable?	In real-time process, the architecture must be open towards addition of new units like smart meters or any other IoT assets.
	What is network coverage?	The minimum expected coverage area is the perimeter of total city. Also underground deployed devices need to connect.
Security	Does the network support end-to-end encryption and authentication?	The design must meet the required security standard to avoid future ambiguities.
Communication infrastructure	Does the network uses licensed spectrum?	It values high quality of service and low quality of interference.
	Is there any need of additional infrastructure?	No need of complex implementation.
	What about if the spectrum provider get changed?	Keep always ready the back up to avoid vendor-lock problem.
	Does the service have scope to start new features?	Always, it is better to have such service provider who updates its service with time.
	Does the service support open data formats?	To make data portable.
	Does the platform support Big Data?	The database must accommodate huge amount data for analysis purpose and to save history of resource consumption.

smart applications. The integration of technology with conventional water system can solve many glasses of water-related problems such as:

1. reducing the gap between demand and supply of hygienic and efficient water treatment and distribution;
2. forecasting and predicting natural or man-made alarming situations;
3. enhancing the reliability of water distribution network (WDN) and remote monitoring of water utilization;
4. real-time monitoring of asset condition;
5. real-time water quality monitoring; and
6. real-time water consumption monitoring to support water conservation for the future.

With rapid population growth in cities, the urbanites are facing tough competition in the accessing process. The resources such as electricity, natural gas, and water need to be managed very carefully to ensure future use and sustainability [9]. SW is a new paradigm with ICT convergence for future generations. The ICT-based SW is expected to be helpful to the mankind to deal with various contexts such as a scarcity of water, maintaining water quality, and proper distribution of water according to the situation demand [8]. In order to improve the quality of life, reduce the effects of

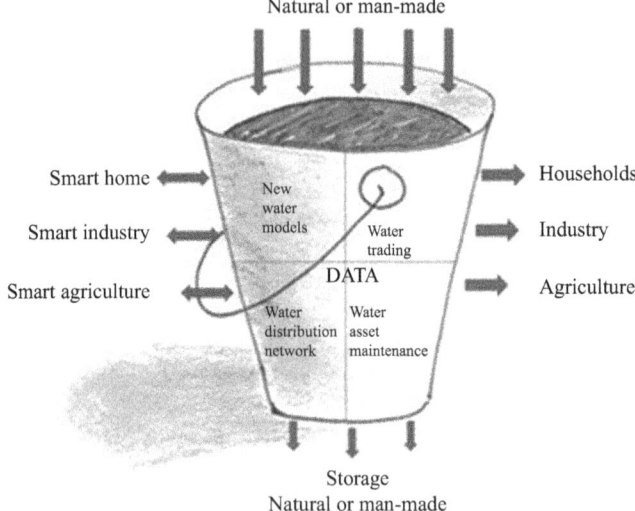

Figure 3.3   Water supply process

human activities and the way of water utilization new technologies are under search. People living in developed and underdeveloped countries are facing problems like availability of clean water and water at the right time. To meet the demand of clean water by the growing population and urbanization has become a challenge at present. During the process of addressing this issue, it invites many parallel challenges such as the cost of management, water transmission, storage, distribution, and billing for the consumption of water. The lifestyle of the people is changing rapidly and at the same time, the capacity of the people to pay has also increased. These things have certain negative effects on the use of water [10]. In Figure 3.3, the authors presented a water supply process in the digital city according to their own understanding.

3.3   SW challenges

The deployment process of an SWS is very tough because of several challenges offered by our conventional society [17]. Here, the challenges are segregated into two broad categories, i.e. civil challenges and technical challenges. The source of *civil challenges* is mainly the government and people. The challenges are the following:

1. uncontrolled population [18];
2. rapid urbanization [18];
3. unawareness among people regarding the value of water and its proper usage [18];
4. lack of fund to modernize the existing conventional WDN;
5. lack of interest in governing bodies [19,20]; and
6. lack of advertising regarding waking up people to adopt new technologies [19,20].

During the deployment phase, the following *technical challenges* are encountered:

1. low-level maturity in ICT infrastructure standardization for three primary components such as sensors, IoT, and data analytic tools;
2. lack of experts to handle modern tools; and
3. lack of fund and interest to do further research on SWS.

This section covers three key aspects of SWS, in relation with both the key operational aspects of WDN asset management, leak management, and the SW metering at customer's end. For each aspect, this chapter focuses on the challenges, objectives, and current technologies.

3.3.1 Asset maintenance

Globally, water loss through damaged or degraded assets like a water tap, water tank, and water pipes remain a top need for utilities looking to make investments in new innovations [8]. System proprietors and administrators are progressively looking to decrease dependence on conventional technologies and procedures, such as pipe replacement or system clearing, with increasingly complex, smart arrangements including imparting sensors and data analytics [21]. In a move towards this plan of asset management, many countries have been adopted several innovative methods. For example, in 2007, in the UK, the south-west water executed a plan with 4,000 dNet sensors, 250 pressure reduction valve condition monitoring loggers, and i20's iNet platforms from UK-based pressure management specialist i20 [22]. The goal of this operation is to monitor valve condition, flow, pressure, and transients across its network while optimizing energy efficiency for pumping operations. Likewise, several SW projects are in the process around the world [23].

3.3.2 Leakage maintenance

The current water supply and distribution network is mainly dependent upon pipes, with these pipes normally buried underground, and it's the prime challenge to detect and fix leakages in the pipe [13]. Leakages detection and prevention is one of the prime acts in WDN towards SW project. The main reason for leakages in pipes is water pressure, when the pressure of supplied water is more than the pressure of the pipe, set by the manufacturer than the water starts getting deformation and explode the pipelines which cause leakages in pipelines.

3.3.3 SW metering maintenance

In today's global water scarcity situation, the deployment of smart water meter (SWM) architecture is very much essential to monitor water usage by household and industries. The SWMs are normally deployed in customer premises to the consumption rate based on usage. This helps to customer to gauge and to adopt preventive actions such as reducing consumption and detecting leakages and customize their usages. The major problem which lies with smart metering is a fault in the communication

process as smart meter need to communicate with the utility centre for report generation. Hence, regular maintenance is very much essential to avoid ambiguity between
customer and resource provider.

3.3.4   Challenges

The major observed challenges during the deployment phase of the SW application
are the following:

- Low cost: The wide area deployment demands a large-scale investment on
  infrastructure which included the cost of equipment and installation charges.
- Miniaturization: The physical deployment of intelligent sensing technologies
  needs to be very compact with the integration of integrated circuits.
- Scalability: The rapid population introduces many new colonial structures which
  demand the deployment of much new hardware and integration with base
  software.
- Flexibility: The underground and remote deployment creates hindrance in the
  fault identification process.
- Modular design: To bring user convenience in the new model, the modular design
  is very much essential.
- Reliable communication: The machine-to-machine (M2M) communication environment essentially need to back up with auto-fault recoverable technology for
  uninterrupted communication.

3.4   Proposed structure for SW

In the previous section, we came to know the various challenges involved in the
process of digitalization. In the current section, we will incorporate those insights to
model a sustainable SWS. In Figure 3.4, the resultant model for sustainable SWS is
illustrated. The SWS can be described by seven layers of enablers:

1. natural,
2. stakeholder and their involvement,
3. policies of government and locality,
4. goals,
5. infrastructure,
6. ICT, and
7. resilience.

Layer 1: The natural weather and environment and natural resources are defined as
   a natural enabler. Nature is the foundation of all life either it is urban or rural
   hence it needs to be preserved accordingly. SW process will be successful if
   the proposed solutions will allow recycling and controlled extraction of natural
   resources, which ensures natural regeneration and avoids harmful emissions [24].

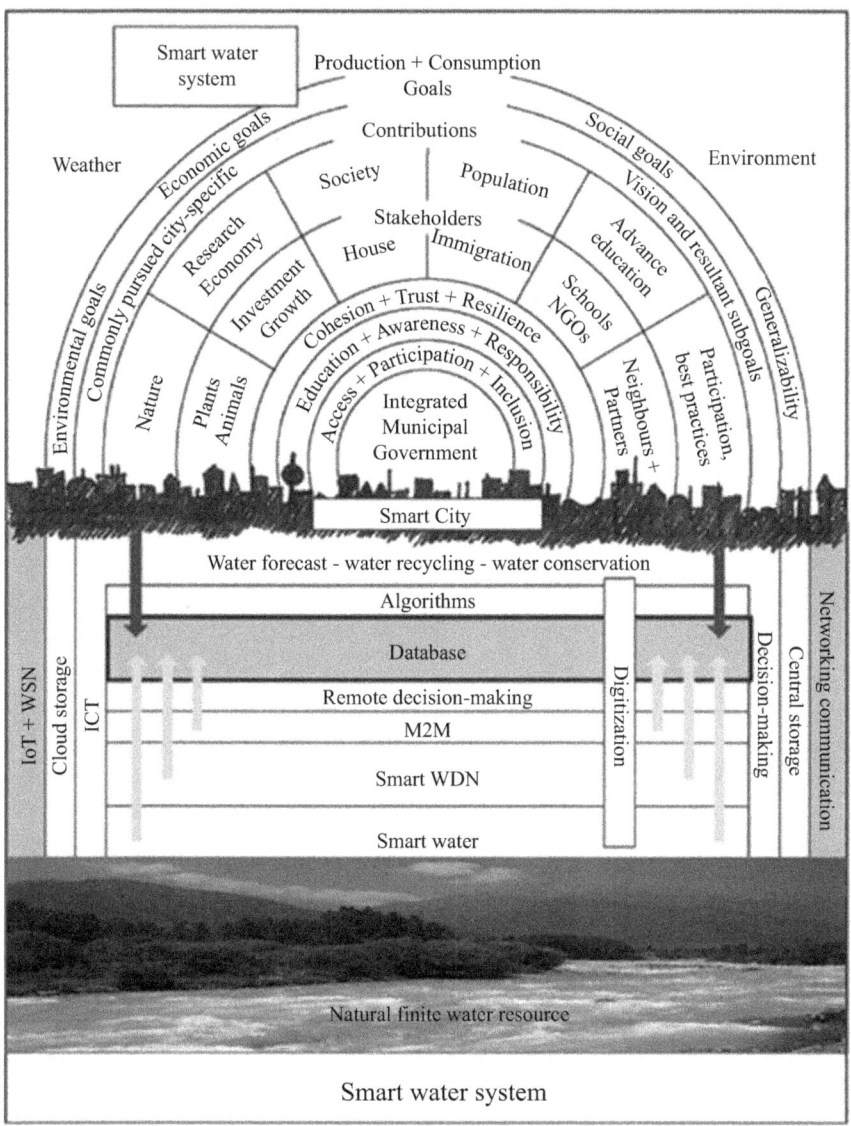

Figure 3.4   Proposed SW model

Layer 2: The second layer of the proposed model describes stakeholders (human) and their involvement. In the introduction section, it was elaborately explained that the needs of urbanites bring a shape to the smart city. Therefore, it is important for government and municipality to make policies which are human-centric. To bring awareness among people, the education system needs to measure against practical implementations. The smart city deserves smart people to maintain the smartness. Through smart citizen only can bring synchronization between demand and supply which avoids unnecessary wastage of natural resources like water, soil, etc. [25,26].

Layer 3: The third layer meant to bring urban governance where the policies and rules and regulations must be framed by the integration of government and municipality as per urbanites. The purpose of such integrated governance is to initiate several plans and activities by considering the various affairs of public, private, and civilian sectors. The state government is responsible for representing its national interest and the framework. In doing so, it actively coordinates the other five layers by providing proper priority to the city's objective and convenience of urbanites [27]. Through the integrated model, we can achieve procedural, organizational, and structural demand of urban society. It strives to achieve efficiency, sustainability, and transparency in inter-sectoral and interdisciplinary plans which ultimately produce economic growth, proper resource utilization without wastage, and neighbours partnership.

Layer 4: The goal of the city can be achievable only and when the planning becomes a joint venture by citizens and the government. The joint venture process brings proper shape in the decision-making process which raises common goal and mission [28].

Layer 5: The fifth layer, i.e. infrastructure, is one of the core pillars of the smart system (e.g. SW, grid, parking, and healthcare) [29]. An appropriately designed, interconnected, flexible, and resource-efficient infrastructure only can justify the integration of modern technologies and their functionalities [30]. The integrated model infrastructure enables to get information, resource sharing, and central control. The very suitable example of such infrastructure is zero-water wastage building at Bhubaneswar under smart city project [31].

Layer 6: ICT is a sixth and advanced layer which is formed in the combination of urban process and infrastructure. It influences all layers except the natural one. The interconnected infrastructure with ICT is able to gather data from all around corners of an urban city, and in addition, it provides data security and privacy.

Layer 7: In the process of digitization, the city is always open to vulnerability either for natural or man-made situations. The seventh layer, i.e. resilience, is an indeed functionalities which ensure the sustainability by providing fault tolerance and recovery process and by avoiding the fault occurrences [32].

3.5 IoT as solution

The term *assets* in water context means several types of equipment such as pipeline, valve, tap, and storage tank. The continuous contamination of water with these object starts some sort of reactions, which introduces many critical situations over the period of time. The handling of such unwanted reactions is the major challenge for utility owner against ageing technologies. In this section, a few approaches related to SW solution have been explored.

3.5.1   IoT in asset maintenance and leakage detection

Before discussing the solution based on IoT, we first need to understand, What is IoT? The term IoT stands for the IoT is a system where all participating objects such as computing devices, mechanical and digital devices, animals, or people, etc., can communicate with each other. Here, each participating device is assigned with a unique identifier with the ability of human-to-human or human-to-machine communication. In the smart city project, IoT is an inevitable technology. Figure 3.5 illustrates the role of IoT in SW objective.

The various operations with water demand the use of several types of assets, for example, the following: for storage, there is a need of water tank; for distribution, there is a need for pipeline; for water flow control, there is a need for valve; likewise, many devices have been required in a different context of water processing [25]. It is well known to all that the water is the combination of two elements that are hydrogen and oxygen (H_2O) whereas, the other assets also have their own chemical composition like pipeline normally made up of plastic or cast iron, valves mainly made up of with iron or steel, etc. When these chemical compositions become in contact with each other for a long period of time that leads to a new chemical reaction and the resultant comes in the form of unhygienic water quality. There are several

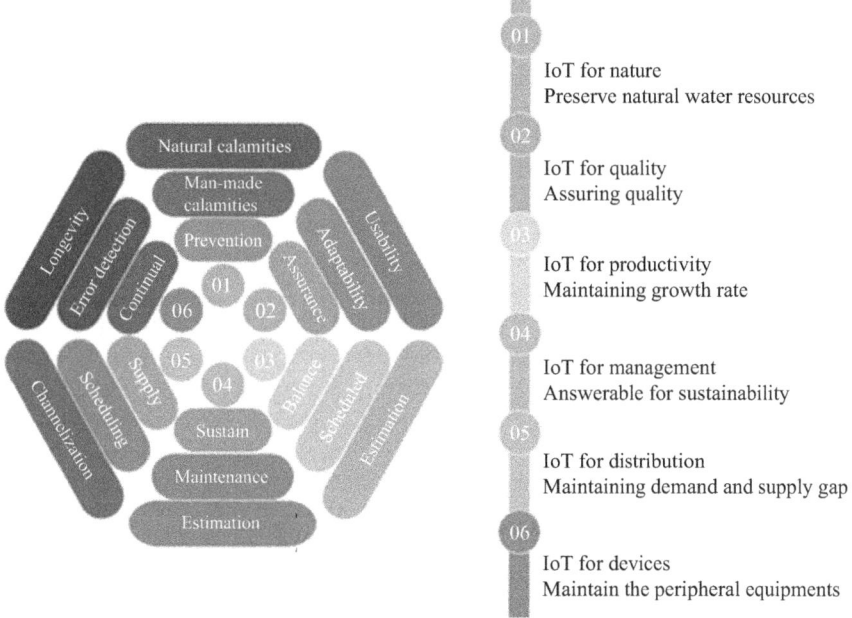

Figure 3.5   IoT involvement in SW

research papers published in the past which highlights plastic and iron quality as per drinking standard.

The advancement in technology like IoT, ICT, and Wireless Sensor Network (WSN) tremendously helps to solve these problems from the remote end [33]. For example, Libelium launched an SW wireless sensor platform to simplify remote water quality monitoring [34]. Equipped with multiple sensors that measure a dozen of the most relevant water quality parameters, Waspmote Smart Water is the first water quality-sensing platform to feature autonomous nodes that connect to the cloud for real-time water control. Waspmote Smart Water is suitable for potable water monitoring, chemical leakage detection in rivers, remote measurement of swimming pools and spas, and levels of seawater pollution. The sensor can also be deployed in a water tank to gauge the water level and switch off power automatically based on the sensor signal. Same way in the water pipes also sensor can be deployed which can gauge pressure in the pipe and if any diversion found then that cab reported with longitude and latitude of the leakage point for fixing purpose. There are many ways and approaches for handling the water-related problems with the help of IoT and sensors.

3.5.2 IoT in SW metering

SW metering is a system that collects, measures, and tracks distribution and consumption and communicates with utility centre. It is a virtual aspect of SW metering which enhances the remote communication regarding WDN. The major functions of communication protocol are data format, data gathering procedure, and decision system based on data mining with big data. The goal is to bring a better surveillance system which will help to track the gap between demand and supply of water. Many research papers have been published in the past to provide automatic metering infrastructure for various applications such as gas, electricity, and water [35].

3.5.3 IoT in SW bottle

The role of IoT is no more limited with big appliances rather it also can be integrated with small day-to-day usable products like a water bottle (e.g. DrinKup bottle). The core properties of this bottle are the following: it is eco-friendly, Bluetooth 4.2 enabled, and bisphenol A free. The battery of this bottle lasts for 30 days. The core objective of this IoT-enabled bottle is to acknowledge the consumer about its dehydration state, and additionally, it also informs about the temperature of water like if the water is too hot or cold. This smart technology rests upon a smart lid which is occupied with an ultrasonic sensor which keeps track of water intake throughout the day [36]. This smart device is supported by an app for the user interface.

3.5.4 IoT in SW distribution network

Another dimension of IoT-called predictive analytics is one of the major domains which help to understand the future demand of water for a particular city. This is done by keeping record of the water usage amount in a historical manner. This historical data when analysed through predictive analytics by considering several other factors such as weather condition of the city, special events and holidays, and population the approximate value can be predicted in the pre-hand state [37].

3.5.5   IoT in SW irrigation

The country like India, where irrigation is the backbone of the country economy, seriously needs an SW irrigation model to avoid water wastage. The global warming deviously influenced the natural raining process so no more the farmer can be dependent on natural raining for cultivation. Here, the IoT plays an important role to measure the water demand level. The irrigation process normally happens at a particular time for a stipulated duration. During such a case, the availability of water can be assured by using IoT-enabled mechanisms. The whole process is dependent upon various kinds of sensors like soil sensor and a weather sensor. These sensors communicate with each other through a central server for decision-making. This mechanism works on an automated schedule level which decides which zone need how much water for which crop. This helps to conserve extra wastage of water during the irrigation process [37]. The case study for this can be studied in [38].

3.5.6   IoT in remote controlling of SW assets

IoT also helps to monitor the peripheral devices like pump and valve which are associated with water supply. The scheduling process through IoT enables to monitor (shut-down/switch-on) these devices from the remote end. It also helps to broadcast the future water supply rate to the residents which reduce the burden of water management from both user and utility centre end. That helps to manage the water resource in an adequate way especially during the water-shortage period [37].

3.5.7   IoT for SW pricing or revenue generation

The smart revenue generation from water energy is also possible by the use of IoT. The predictive analysis can be used to determine the water pricing at various times of the day. This will bring a moderate consumption pattern among consumers, and in parallel, it will generate revenue for the government [37].

3.5.8   IoT for smart billing for SW

The deployment of smart meters for gauging water consumption can be extended with smart billing module. The traditional water billing process needs human to intervene to record water consumption from the water meter. In many cases, it has been observed that meters are deployed within the building of consumers which invites many obstacles for meter reader to get or collect those data. The smart meter can communicate the last reading to a central point from where the meter reader can easily get the data for billing purpose. The IoT-enabled devices have such a communication model which helps to generate the bill form remote end.

3.5.9   IoT for measuring deep water quality

The water resource is not only restricted to the human race. It is also our responsibility to maintain the water quality level for water habitats. In this line of thought, there are several projects running to assure that the lifestyle of water habitats should not get disturbed by any means. Few measure case studies in this line of thought are as follows:

* Protecting and conserving the beluga whale habitat in Alaska with Libelium's flexible sensor platform [39].
* Controlling fish farms water quality with smart sensors in Iran [40].
* Preserving endangered freshwater mussels in the Ohio River with an SW project [41].

3.5.10   IoT for predicting water-based natural calamities

The improper infrastructure and overpopulation growth invite two types of disasters that are man-made calamities and natural calamities. The use of IoT and ICT in this sector brings a reliable and stable structure in terms of accuracy and communication. The deployed sensors in this kind of project primarily record water flow rate and its behaviour. In this regard, two major projects [42,43] can be referred to understand the deep involvement of IoT for disaster management in terms of early flood detection and generating an alarm.

Here, in Table 3.2 few ongoing SW projects have been illustrated [44].

3.6   Lessons

In this chapter, an IoT-based SW structure is demonstrated where different layer-based enablers have been discussed. The SW concept is an emerged trend with the integration of IoT and convergence of ICT solutions. The proposed structure says

Table 3.2   SW projects

Country	Project	Objective
Bangalore, India	Protecting Lakes from Land Grabbers	Awareness spreading among people for protecting the lakes
Cape Town, South Africa	Water Conservation Program	Persuading people to use less water, deploying latest technologies to use water efficiently
Sorocaba, Brazil	Cleaning Up Polluted Rivers	
Beira, Mozambique	Tackle the urgent problem of Urban Strom Water	Preventing soil erosion, flooding, and threat to infrastructure and life
Lima, Peru	Promote Water Saving	Promoting cultivate willingness to pay proper water tariffs, and aware people about environment issues

that a strategic and structural approach has to be adopted in order to address several glasses of water-related challenges. The following lessons can be concluded in this chapter.

- The deployment of SW is a systematic and evolutionary process which demands the involvement of emerging technologies and a long period of time.
- The technological deployment must be within the boundary of nature's law.
- Modern education and awareness programs are indeed to walk in parallel with technologies.
- Resilience and scalability are the two main features of a smart city.
- In comparison to all smart cities applications, the non-renewable source-based application needs serious attention.
- The cost of new IoT-enabled devices for SW must be within the range of consumer.

3.7   Conclusion

In this chapter, a novel IoT-based SW layer-based model is discussed which focuses on SW management through seven layers. The very principle of every advancement is helping society provided, without disturbing the nature's law. In the proposed structure, this principle is carefully considered, and the same time, it's also taken care of the efficient utilization of natural resources. Each of the layers is highlighting the various factors which are influencing the SW decision. The water is a basic need of every living being on earth and unfortunately, till date, the science is incapable to bring out an alternative method to produce water. Hence, this non-renewable energy needs to be invested very carefully so that the future generation will not suffer from water scarcity. At the same time, the other side of science is enough matured to provide smart solutions to manage the water resource and to avoid water wastage. Personally, we feel the coverage area of the WDN is so large that it is very difficult for a human to manage it efficiently. Hence, emerging technologies like IoT, ICT, and WSN could be the solution. This chapter will help novice researcher of water domain to understand the insights of water-related problems and challenges.

References

[1] Al-Turjman, F., and Malekoo, A., 2019. Smart parking in IoT-enabled cities: A survey. Sustainable Cities and Societies, vol. 49, 101608, ISSN 2210-6707, https://doi.org/10.1016/j.scs.2019.101608

[2] Al-Turjman, F., and Abujubbeh, M., 2019. IoT-enabled smart grid via SM: An overview. Future Generation Computer Systems, vol. 96, no. 1, pp. 579–590.

[3] Al-Turjman, F., 2019. 5G-enabled devices and smart-spaces in social-IoT: An overview. Future Generation Computer Systems, vol. 92, no. 1, pp. 732–744.

[4] http://smartcities.gov.in/

[5]   Desouza, K.C., Swindell, D., Smith, K.L., Sutherland, A., Fedorschak, K., and Coronel, C., 2015. Local Government 2035: Strategic Trends and Implications of New Technologies. Issues in Technology Innovation, Number 27, May 2015.

[6]   Al-Turjman, F., 2019. The road towards plant phenotyping via WSNs: An overview. Computers & Electronics in Agriculture, vol. 161, pp. 4–13.

[7]   Al-Turjman, F., Altrjman, C., Din, S., and Paul A., 2019. Energy monitoring in IoT-based ad hoc networks: An overview. Computers & Electrical Engineering Journal, vol. 76, pp. 133–142, 2019.

[8]   Saputro, N., Akkaya, K., and Uludag, S. A survey of routing protocols for smart grid communications. Computer Networks, vol. 56, no. 11, pp. 2742–2771, ISSN 1389-1286, https://doi.org/10.1016/j.comnet.2012.03.027.

[9]   ARD (2014, 23 March) Spanien/Marokko: Der tdliche Zaun von Melilla [Video]. http://www.daserste.de/information/politik-weltgeschehen/weltspiegel/videos/spanien-marokko-der-toedliche-zaun-von-melilla-100.html. Accessed 16 Jan 2015.

[10] ACTCOSS, CCSERAC, 2003. Saving our water resources and equitable and sustainable policy for the ACT. ACT Council of Social Services and Conservation Council of the South East Region and Canberra (Canberra).

[11] Bagozzi, R.P., and Yi, Y., 1988. On the evaluation of structural equation models. Journal of the Academy of Marketing Science, vol. 16, pp. 74–94.

[12] Barrett, G., 2004. Water conservation: The role of price and regulation in residential water consumption. Economic Papers: A Journal of Applied Economics and Policy, vol. 23, no. 3, pp. 271–285.

[13] Birrell, R.J., Rapson, V.J., and Smith, T.F. 2005. Impact of Demographic Change and Urban Consolidation on Domestic Water Use. Melbourne Vic Australia: Water Services Association of Australia.

[14] http://www.libelium.com/preserving-endangered-freshwater-mussels-in-the-ohio-river-with-a-smart-water-project/.

[15] https://theurbantechnologist.com/2013/07/14/smarter-city-myths-and-misconceptions/.

[16] https://en.wikipedia.org/wiki/$Cyclone_F ani$.

[17] Gilg, A., and Barr, S., 2006. Behavioral attitudes towards water saving? Evidence from a study of environmental actions. Ecological Economics, vol. 57, pp. 400–414.

[18] Britton, T., Stewart, R.A., and O'Halloran, K., 2009. Smart metering providing the foundation for post meter leakage management. 25th to 28th October 2009. IWA Specialist Conference – Efficient 2009, Sydney, Australia.

[19] Corral-Verdugo, V., Carrus, G., Bonnes, M., Moser, G., and Sinha, J.B.P. (2008). Environmental beliefs and endorsement of sustainable development principles in water conservation: Toward a new human interdependence paradigm scale. Environment and Behavior, vol. 40, no. 5, pp. 703–725. https://doi.org/10.1177/0013916507308786.

[20] CSIRO, 2002. Perth domestic water-use study household ownership and community attitudinal analysis. NSW, Australian Research Centre for Water in Society CSIRO Land and Water.

[21] Gleick, P.H., 1996. Basic water requirements for human activities: Meeting basic needs. Water International, vol. 21, no. 2, pp. 83–92.

[22] Heinrich, M., 2007. Water End Use and Efficiency Project (WEEP) – Final Report. BRANZ Study Report 159. Branz, Judgeford, New Zealand.

[23] Giurco, D., Carrard, N., McFallan, S., *et al.*, 2008. Residential End-use Measurement Guidebook: A Guide to Study Design, Sampling, and Technology. Prepared by the Institute for Sustainable Futures, Victoria.

[24] Kim, S.H., Choi, S.H., Koo, J.K., Choi, S.I., and Hyun, I.H., 2007. Trend analysis of domestic water consumption depending upon social, cultural, economic parameters. Water Science and Technology: Water Supply, vol. 7, no. 5–6, pp. 61–68.

[25] Savenije, G.H.H., and van der Zaag, 2002. Water as an economic good and demand management: Paradigms and pitfalls. Water International, vol. 27, no.1, pp. 98–104.

[26] WHO, 2005. Minimum Water Quantity Needed for Domestic Use in Emergencies. WHO-Technical Notes for Emergencies.World Health Organization, Switzerland.

[27] https://www.waterworld.com/articles/wwi/print/volume-33/issue-2/technolog y-case-studies/smart-water-what-to-expect-in-018.html.

[28] Worthington, A.C., and Hoffman, M., 2008. An empirical survey of urban water demand modeling. Journal of Economic Surveys, vol. 22, no. 5, pp. 842–871.

[29] Stewart, R.A., Willis, R., Giurco, D., Panuwatwanich, K., and Capati, G., 2010. Web-based knowledge management system: Linking smart metering to the future of urban water planning. Australian Planner, vol. 47, no. 2, pp. 66–74.

[30] Middlestadt, S., Grieser, M., Hernandez, O., *et al.*, 2001. Turning minds on and faucets off water conservation education in Jordanian schools. The Journal of Environmental Education, vol. 32, no. 2, pp. 37–45.

[31] http://www.libelium.com/newsletter/

[32] Misiunas, D., Lambert, M., Simpson, A., and Olsson, G., 2005. Burst detection and location in water distribution networks, Water Science and Technology: Water Supply, vol. 5, no. 3–4, pp. 71–80.

[33] https://iiot-world.com/connected-industry/smart-water-management-using-internet-of-things-technologies/.

[34] White, S., Turner, A., Fane, S., and Giurco, D., 2007. Urban water supply-demand planning: A worked example. 4th IWA Specialist Conference on Efficient Use and Management of Urban Water Supply, Jeju, Korea.

[35] Public Utilities Board Singapore, 2016. Managing the water distribution network with a Smart Water Grid. Smart Water 1, 4. doi:10.1186/s40713-016-0004-4.

[36] Al-Turjman, F., and Alturjman, S., 2018. Context-sensitive access in industrial Internet of Things (IIoT) healthcare applications. IEEE Transactions on Industrial Informatics, vol. 14, no. 6, pp. 2736–2744.

[37]   https://www.mbreviews.com/best-smart-water-bottle/.
[38]   http://www.libelium.com/controlling-fish-farms-water-quality-with-smart-sensors-in-iran/.
[39]   Mohapatra, H., and Rath, A.K., 2019. Detection and avoidance of water loss through municipality taps in India by using smart tap and ICT. IET Wireless Sensor Systems, 2019, 9, (6), pp. 447–457, DOI: 10.1049/iet-wss.2019.0081, IET Digital Library, https://digital-library.theiet.org/content/journals/10.1049/iet-wss.2019.0081.
[40]   http://www.libelium.com/protecting-and-conserving-the-beluga-whale-habitat-in-alaska-with-libeliums-flexible-sensor-platform/.
[41]   http://www.libelium.com/early-warning-system-to-prevent-floods-and-allow-disaster-management-in-colombian-rivers/.
[42]   http://www.libelium.com/controlling-quality-of-irrigation-water-with-iot-to-improve-crops-production/.
[43]   http://www.libelium.com/early-flood-detection-and-warning-system-in-argentina-developed-with-libelium-sensors-technology/.
[44]   https://www.theguardian.com/global-development-professionals-network/2016/feb/29/five-of-the-best-water-smart-cities-in-the-developing-world.
[45]   Fujii, Y., Yoshiura, N., Takita, A., and Ohta, N., 2013. Smart street light system with energy saving function based on the sensor network. e-Energy'13.
[46]   https://en.wikipedia.org/wiki Global Internet usage.

Chapter 4
Contiki-OS IoT data analytics

Muhammad Rafiq[1], Ghazala Rafiq[1], Hafiz Muhammad Raza ur Rehman[1], Yousaf Bin Zikria[1], Sung Won Kim[1], and Gyu Sang Choi[1]

Advancements in sensor network have evolved rapidly in recent years, and devices are smart enough to build and manage their network and route optimization referred to as Internet of Things (IoT). Numerous IoT operating systems (OSs) are developed for resource-constrained IoT devices. Contiki IoT OS is a widely used IoT OS by researchers and practitioners. Contiki-OS Cooja emulator is recognized as one of the favourite tools of researchers for running large-scale simulations and observing the results before the real-time deployment. Cooja generates execution logs for all the activities of the network simulation. However, there are no tools or programs available to summarize and analyse the big log files generated by Cooja. This slows down the research pace for complex network scenarios and makes it difficult to compare with existing bench marks and research work. In order to help researchers, an evaluation tool which gathers information, analyses and develops simulation log results is required. It provides detailed individual mote statistics as well as complete IoT network statistics. In this chapter, we discuss three algorithms and their merits and demerits. First, the proposed scheme scans the generated log file and provides summary of all the IoT motes in separate files. This technique is useful for very large files and complex operation, although it requires more hard disc space for temporary files. Second, the proposed algorithm scans log file to summarize data in memory. This algorithm requires additional space for temporary files and scans source files many times consequently, and it requires more time to complete the evaluation. Third algorithm scans log file exactly once, does not require any additional space for temporary files and computes summaries in memory. It makes processing really fast, and can work without temporary files generated. All three algorithms are helpful in different IoT deployment scenarios; therefore, researcher can choose according to their preference of memory requirements, file sizes, and time constraints.

[1]Department of Information and Communication Engineering, Yeungnam University, South Korea

4.1   Introduction

The IoT [1,2] is a network of smart objects with sensors, embedded electronics and software connected with each other using self-maintained wireless network. Things or smart objects collect physical data and exchange with other devices using the network. More often IoT devices are controlled from a remote management computer and updated dynamically with new instructions or more modules to handle different or enhanced functionalities. It makes it possible to connect physical world with smart devices [3,4]. Smart spaces in Social IoT [5], IoT-enabled smart grid [6], smart parking [7], security and privacy in smart cities [8] and mobile couriers in smart cities [9] are some recent research topics in this field.

IoT OS availability with support of latest standards and heterogeneous variety of motes are crucial for next generation IoT. Scarcity of resources on IoT motes puts strict requirements on IoT OS overall code size. In addition, IoT OS needs to provide all the desired layered functionalities to be part of large-scale IoT network. Hence, IoT OS requires to be highly efficient and is optimized to manage the IoT mote resources [10].

Data gathered by various means are mainly the raw information. Data size is increasing drastically day by day. Hence, it is becoming difficult to efficiently store, analyse and get meaningful results [11,12]. Consequently, it helps in decision-making. This process is known as data analysis. Data analytics techniques and technologies are very helpful to analyse large amount of data sets and extract information for future direction. Well-known approaches for data analysis make it different from traditional kind of analysis. It includes programmatic approach, i.e. using any programming language to deal with large-scale data. Second approach can be data-driven. In this approach, some hypothesis is used to analyse the data. It is also called the hypothesis-driven approach. Third approach is iterative. In this technique, data is divided into bunches or groups and analysis is carried out iteratively until the desired results are not achieved. Another accurate big data handling approach is to use data attributes. In the past, analysts dealt with hundreds of attributes or characteristics of the data sources; however, now there are thousands of attributes and millions of observations. Data generated by Contiki-OS Cooja simulator are structured and deterministic. Analysing Contiki-OS log messages is a time-consuming process to get the overall network performance. A specialized analytics tool can generate a precise summary that helps researchers to quickly compare and benchmark the research scenarios.

Data are increasing gigantically day by day. Researchers are in need for certain data analytic tools. These tools should make it convenient for researchers to examine/analyse the data in the shortest possible time without compromising the evaluation criteria and quality. To achieve this goal, one of the possibilities is to apply computerized script to raw data that can generate some useful information out of it.

Simulations performed in Contiki-OS Cooja simulator generate execution logs for all the activities of the network simulation. However, there are no tools or programs available to summarize and analyse the big log files generated by Cooja. This slows down the research pace for complex network scenarios and makes it difficult to compare with existing bench marks and research work. In order to help researchers, we have gathered information and developed three algorithms to analyse Contiki log files. It provides detailed individual mote statistics as well as complete IoT network statistics.

Cooja-generated execution log file is processed by a processing script that is developed using Python programming language. This processing script goes through the whole generated log file line by line. Afterwards, it provides detailed individual mote and network-level performance metrics. First, the proposed scheme scans the generated log file, creates temporary files on the disc during evaluation and provides summary of all the IoT motes and individual mote performance in an output text file. This technique is useful for very large files and complex operation, although it requires more hard disc space for temporary files. Second, the proposed algorithm scans log file to summarize data in memory. This algorithm requires additional space for temporary files and scans source files many times consequently. It also requires more time to complete the evaluation. Third, the algorithm scans log file exactly once, does not require any additional space for temporary files and computes summaries in memory. It makes processing really fast, and can work without temporary files.

This chapter is structured as follows. Section 4.2 provides summary of Contiki OS. Section 4.3 elaborates Contiki-OS different message types and structures. Section 4.4 explains the proposed algorithms along with pseudocodes and flow charts. Section 4.5 enlists all the metrics used for mote and network statistics. Section 4.6 contains results and discussion for the proposed algorithms. Finally, Section 4.7 concludes the chapter.

4.2 Contiki OS

Dealing with IoT, one of the best available options for OS is Contiki OS [13]. Contiki is an open source OS and provides comprehensive standard features for rapid development and deployment. It connects small low-power and low-cost IoT devices to the Internet. This OS can be utilized to build complex IoT networks.

Main features of Contiki OS [14] include the following:

- Contiki is an open source OS. It can be utilized in commercial and non-commercial systems.
- Powerful low-power Internet communication can be done using Contiki. It fully supports Internet Protocol version 6 (IPv6) standard and Internet Protocol version 4 (IPv4), along with IPv6 over low-power wireless personal area network (6LoWPAN), Routing protocol for Low-power and Lossy Network (RPL) [15], Constraint Application Protocol (CoAP) and the recent low-power wireless standards. Taking into account the working methodology of ContikiMAC and sleepy routers, even the wireless routers can be battery-operated.
- Fast and easy development is one of the key features of Contiki OS. All applications are coded in standard C language. A vast range of IoT motes with low power are supported by Contiki OS. These wireless devices are easily available.
- Contiki is designed for tiny memory-constrained IoT devices. Therefore, Contiki is highly memory efficient and provides a set of mechanisms for memory allocation.

- A full IP network stack with standard IP protocols is available in Contiki OS. These protocols include User Datagram Protocol (UDP), Transmission Control Protocol (TCP) and Hypertext Transfer Protocol (HTTP).
- Contiki OS also provides energy module for overall power consumption.
- Modules loading and linking at runtime is supported in Contiki.
- Contiki OS uses network emulator/simulator named Cooja. IoT deployment scenario typically contains many motes and on a very large scale. Hence, it is tedious to develop and debug the network on real-time motes. Cooja simulator makes it extremely easy for the developers to develop and debug applications for large-scale networks without actual deployment.
- In IoT network environment, Radio Duty Cycling (RDC) protocols are essential to save energy. ContikiMAC RDC improves the battery efficiency of IoT motes and hence prolongs the network lifetime.
- Contiki OS protothreading provides one of the useful mechanisms to save memory. Protothread is a well-defined combination of event-driven and multi-threading mechanism. It is very useful when waiting for events to occur.
- A network stack called Rime [16] stack provides support for IoT network protocols and facilitates code reuse.
- Regression tests are carried out on regular intervals to ensure the expected working of Contiki-OS code and functionality.
- Applications are compiled using Contiki build system for supported platforms.
- Small chunks of memory are required to run Contiki. Memory required for a typical system to run efficiently is minimal.

Cooja is a cross-layer java-based IoT simulator distributed with Contiki [17,18]. Figure 4.1 shows the typical simulation scenario in Cooja. It allows the simulation of different levels from physical to application layer, and also allows the emulation of supported IoT motes. Emulated nodes are compiled and executed natively. It is similar to how the native platform works. Figure 4.1 depicts in detail the Cooja simulator interface. Emulated nodes use MSPSim and the MSP430 emulator to directly load and execute firmware files. Emulated nodes offer high timing accuracy and source-code debugging. However, it requires more memory and processing power.

Objective function (OF) plays a vital role in the formation of topology in RPL. Figure 4.2 shows working of OF to find the optimized routes in RPL instance. Destination-Oriented Directed Acyclic Graph (DODAG) is constructed using OF [19,20]. It is based on Objective Code Point (OCP). RPL supports upward or downward routing schemes. In order to optimize RPL instance, researchers define new OFs depending upon the objective metric such as energy, expected transmissions (ETX), throughput, packet delivery rate, latency and jitter. Minimum Rank with Hysteresis Objective Function (MRHOF) [21] is the default OF in Contiki. MRHOF is uses ETX to find the optimum route to the destination. Contiki also provides distance vector-based OF called Zero Objective Function (OF0). Contiki also provides support for the user-defined OF.

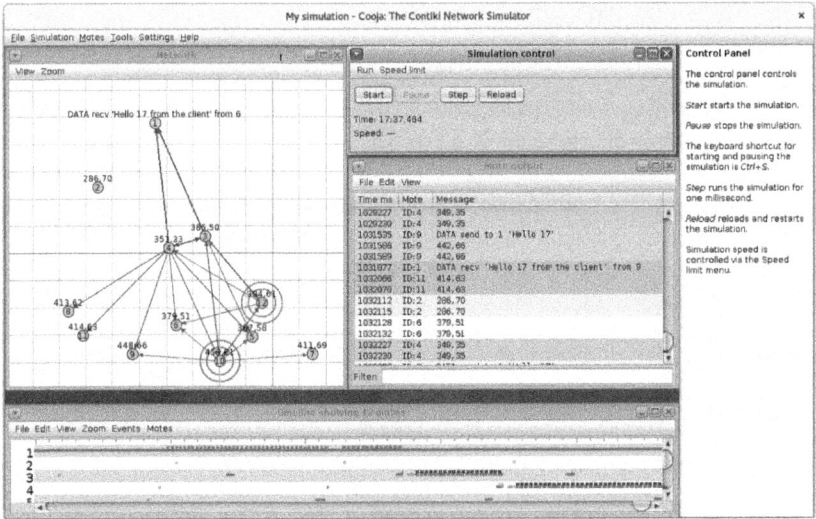

Figure 4.1   Cooja: The Contiki network simulator

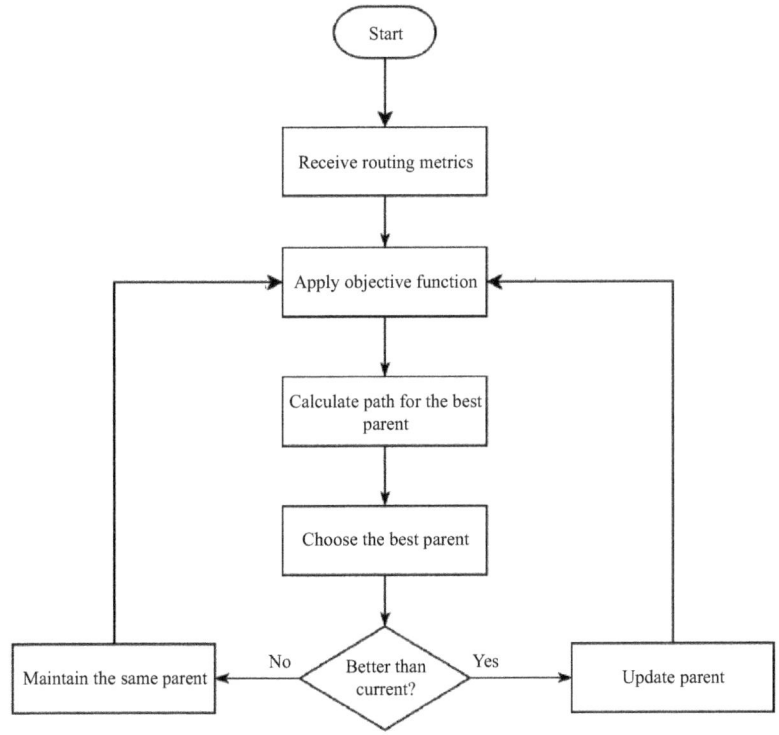

Figure 4.2   Objective function flow chart

4.3   Contiki-OS message structures

Contiki-OS code provides debug messages with conditional compilation switches. Setting DEBUG flag to DEBUG_PRINT enables printing debug log messages. Debug log messages are directed to serial communication; however, while running simulations in Cooja, messages can also be captured by java script for further analysis.

Log messages are printed in output window called mote output window. Separate colours are automatically assigned to different motes. There are a number of messages that are printed in output window. The important messages for scenario under consideration are data messages like SEND, RECV, Media Access Control (MAC) frame drop and directed acyclic graph (DAG) JOIN and control messages like DODAG Information Object (DIO), Destination Advertisement Object (DAO) and DODAG Information Solicitation (DIS). Figure 4.3 represents control messages hierarchy. Figure 4.4 shows send message format. The first field shows the sent time (ms). The second field shows the sender identification number or address, and the next field shows that this message is the sent event. After this, it contains the destination address. At the last, it shows the data message with sequence number. Figure 4.5 shows receive message format. The first field shows the sent time (ms). The second field shows the receiver identification number or address. Next field shows that this

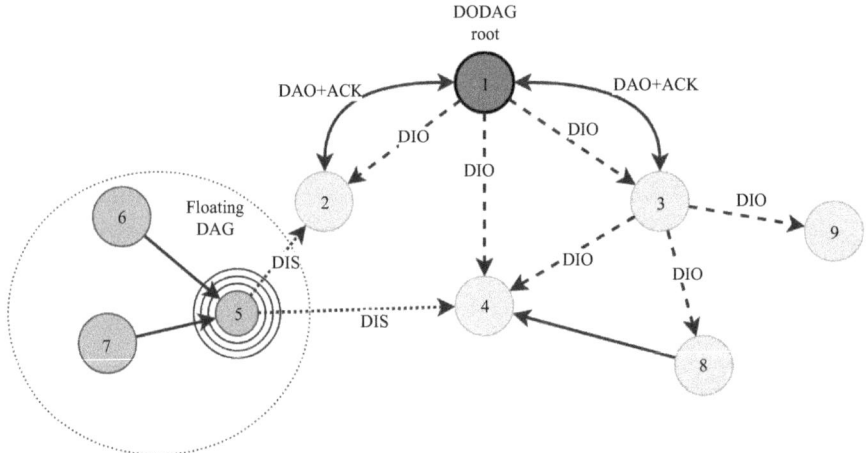

Figure 4.3   RPL control messages

Time (ms)	Sender ID	Message	Destination ID	Data
61,324	ID:2	DATA send	to 1	'Hello 1'

Figure 4.4   Send message format

Time (ms)	Receiver ID	Message	App data	Sender ID
61,324	ID:2	DATA recv	'Hello 1 from the client'	from 2

Figure 4.5   Receive message format

Time (ms)	Mote ID	Message
61,324	ID:2	csma: could not allocate packet, dropping packet

Figure 4.6   Frame drop message

Time (ms)	Mote ID	Type	Message
61,324	ID:2	DIO	fe80::c30c:0:0:2

Figure 4.7   DIO message

Time (ms)	Mote ID	Type	Message
61,324	ID:2	DIS	fe80::c30c:0:0:2

Figure 4.8   DIS message

message is the receive event, and afterwards application data field along with the sequence number. The last field shows the sender address. Cooja generates frame drop messages when it fails to deliver some specific packet to destination. Figure 4.6 explains three parts of the one of the frame drop messages. The first field shows time (ms). Mote ID is the sender node address. The next field is the message itself that indicates MAC frame drop. MAC frame drop count can be used to calculate drop frame percentage at node level as well as at network level. Figure 4.7 shows DODAG DIO message format. The first field shows time (ms) when a DIO message is generated. The second field is the sender node identification. Afterwards, it shows type of message, i.e. 'DIO'. At last, the message part of the structure contains IPv6 address of the node. DIS message plays an important role in the construction of topology. Figure 4.8 shows DIS message format. The first field shows time (ms) when a DIS message is generated. The second field exhibits sender node address. The next field shows type of message is 'DIS'. At the end, it indicates IPv6 address of the sender node. Figure 4.9 shows DAO message format. The first field shows time (ms) when a DAO message is generated. The second field is the node identification number. Afterwards, it shows type of message is 'DAO'. The last field contains IPv6 address of the node.

Time (ms)	Mote ID	Type	Message
61,324	ID:2	DAO	fe80::c30c:0:0:7

Figure 4.9   DAO message

Table 4.1   Node-level metrics

Metric	Formula
Node count	n
Packet count	P
Packet sequence number	p
Packets sent (node)	$P_{sent} = \sum_{p}^{EOF} [T_{sent} > 0]$
Packets received (node)	$P_{received} = \sum_{p}^{EOF} [T_{received} > 0]$
Packet received %	$\dfrac{P_{received}}{P_{sent}} * 100$
Throughput	$\dfrac{P_{received}}{P_{sent}}$
Delay (packet)	$T_{delay} = T_{received} - T_{sent}$
Lost packet (node)	$P_{lost} = P_{sent} - P_{received}$
Delay (node)	$D = \sum_{p}^{EOF} (T_{delay})_p$
Average delay (node)	$D_{mean} = D/P_{received}$
Jitter (node)	$J = \dfrac{\sum_{p}^{EOF} |(T_{delay})_p - (T_{delay})_{p-1}|}{P_{received}}$
MAC frame drop (node)	$P_{drop} = \sum_{p}^{EOF} [drop]$

4.4   Mathematical formulas for required statistics

Table 4.1 shows the metrics used for node-level statistics and Table 4.2 represents network-level metrics. Throughput is a measure of successful number of packets transmitted. It also describes the reliability of the system. It is the ratio between total received packets to the total send packets. MAC frame drop shows the packet drop at the MAC level. This drop occurs if sender fails to deliver the frame to the next hop receiver or a destination. By calculating and analysing MAC frame drops, MAC layer protocol can be optimized to avoid the future packet losses. Packet loss is when one or more packets fail to reach at the destination. Delay shows the time taken by a packet to travel from client to server. In case of delay-sensitive applications, it is very important that average delay of network is known and in bounds. In case of node

Table 4.2   Network-level metrics

Metric	Formula
Packet size	$L = 30$ bytes
Packet sent (network)	$G_{\text{sent}} = \sum\limits_{i=1}^{n}(P_{\text{sent}})i$
Total packet received	$G_{\text{received}} = \sum\limits_{i=1}^{n}(P_{\text{received}})i$
Total network delay	$TotalDelay = \sum\limits_{i=1}^{n}(D)i$
Average network delay	$\dfrac{TotalDelay}{n}$
Jitter	$\dfrac{\sum_{i=1}^{n}J_i}{n}$
Total DIS	$G_{DIS} = \sum\limits_{i=1}^{n}\sum\limits_{p}[DIS]$
Total DIO	$G_{DIO} = \sum\limits_{i=1}^{n}\sum\limits_{p}[DIO]$
Total DAO	$G_{DAO} = \sum\limits_{i=1}^{n}\sum\limits_{p}[DAO]$
TOTAL nodes joined	$G_{join} = \sum\limits_{i}^{EOF}[Joined\ DAG]$
Convergence time	$T_{\text{convergence}} = T[Last\ DAG\ Joined] - T[First\ DAG\ Joined]$
Gross network packets	$G_{\text{total}} = P_{\text{sent}} + G_{DIS} + G_{DIO} + G_{DAO}$
Total network overhead	$OH = G_{DIS} + G_{DIO} + G_{DAO}$
Network overhead %	$\dfrac{OH}{G_{\text{total}}} * 100$

Note: n represents the number of motes in the network

level, it is calculated as average delay of all the received packets. Control overhead is defined as the total number of control packets generated during the network lifetime. It is crucial for the battery-operated IoT network to limit the control overhead to save energy without compromising network reliability. Therefore, control packet overhead can be tuned empirically using overall network control overhead metric. There are three different types of overhead packets DIO, DIS and DAO. Total overhead is equal to the sum of all three control packets. The variance of delay is called jitter. By calculating jitter, the variation in end-to-end delay can be observed. High jitter value corresponds to multimedia streaming quality and consequently degrades the network reliability. It is calculated as by taking difference between two consecutive packet delays for the same node. In case of node level calculation, jitter is taken as average jitter. Convergence time shows how fast all the nodes connected to the server. It is calculated as the difference between time at which first node is connected to time at which last node is connected. Network convergence time highly depends on network size and link quality.

4.5   Proposed algorithms

Data analytics is the key feature of big data as data without proper interpretation is not useful. Simulation log files generated by Cooja cannot be interpreted or analysed without proper processing script. A processing script is developed using Python programming language to calculate node-level metrics as well as network-level metrics. Main idea of Figure 4.10 is to use hard disc drive (HDD) for creating intermediate-level files while calculating different metrics with minimum RAM space. Algorithm 4.1 shows the pseudocode and Figure 4.10 depicts its flow chart. Algorithm 4.1 is divided into four steps for required metrics calculation. Step 1 reads Cooja-generated log file line by line and segregates node-wise data in separate text files. All types of messages related to single node address are appended in its concerned file, e.g. all messages regarding node 2 is in Node2 text file. In similar fashion, messages for all nodes are segregated. In Step 2 of Algorithm 4.1, each node-wise file is opened to search line by line for send, receive or drop event. If line is a MAC frame drop event then a variable specified to calculate node-level MAC frame drop is incremented. If line contains send event then it searches for receive event with that specific sequence number. After getting both send time and receive time for specific sequence number (can also be referred as a packet number), send time, receive time and packet number are appended in a second-level text file, i.e. for Node2 step 2 the filename will be "Node2-2". Repeat the same process for each node and every packet. If *receive* event for a certain packet number is not found, it means the packet is not received and a value of "–1" is set as receive time. In Step 3, the node-level metrics and drop MAC frames are calculated. Results generated for all the nodes are appended in an output file. In Step 4 of Algorithm 4.1, network-level metrics and overheads are calculated and appended in the same output file. All intermediate-level text files are deleted afterwards. It is an efficient approach to divide a large file into chunks, store them on HDD, perform the manipulation according to requirements, generate output and then delete all the intermediate stage files.

Algorithm 4.2 shows the pseudocode and Figure 4.11 depicts its flow chart. It is a closed-outer-loop process. This algorithm uses RAM for processing and storage. Data are analysed line by line. Log files contain data in specific pattern. Based on this pattern, algorithm reads the data line by line; if a node address is not available in current node address list, then it adds in the list. If the node address is already in the list, then it splits the line and checks the message type such as send/DIO/DIS/DAO/MAC. If line contains any control overhead messages, then the value of overhead increases by one. Similarly, it increments MAC drop by one in the event of any MAC frame drop messages. However, if it is a send event, it increments send packets, and stores the sender address, receiver address and message sequence number. After that, the algorithm moves to next line and searches for successful packet reception event. If the packet is successfully received, then it records the received time and calculates packet delay. In case it is not the first packet delay, it also calculates the packet jitter. If the packet is sent but not received, then value of lost packet is incremented. This process continues until the end of log file. Finally, it calculates statistics for the whole network. To accomplish this goal, it first calculates

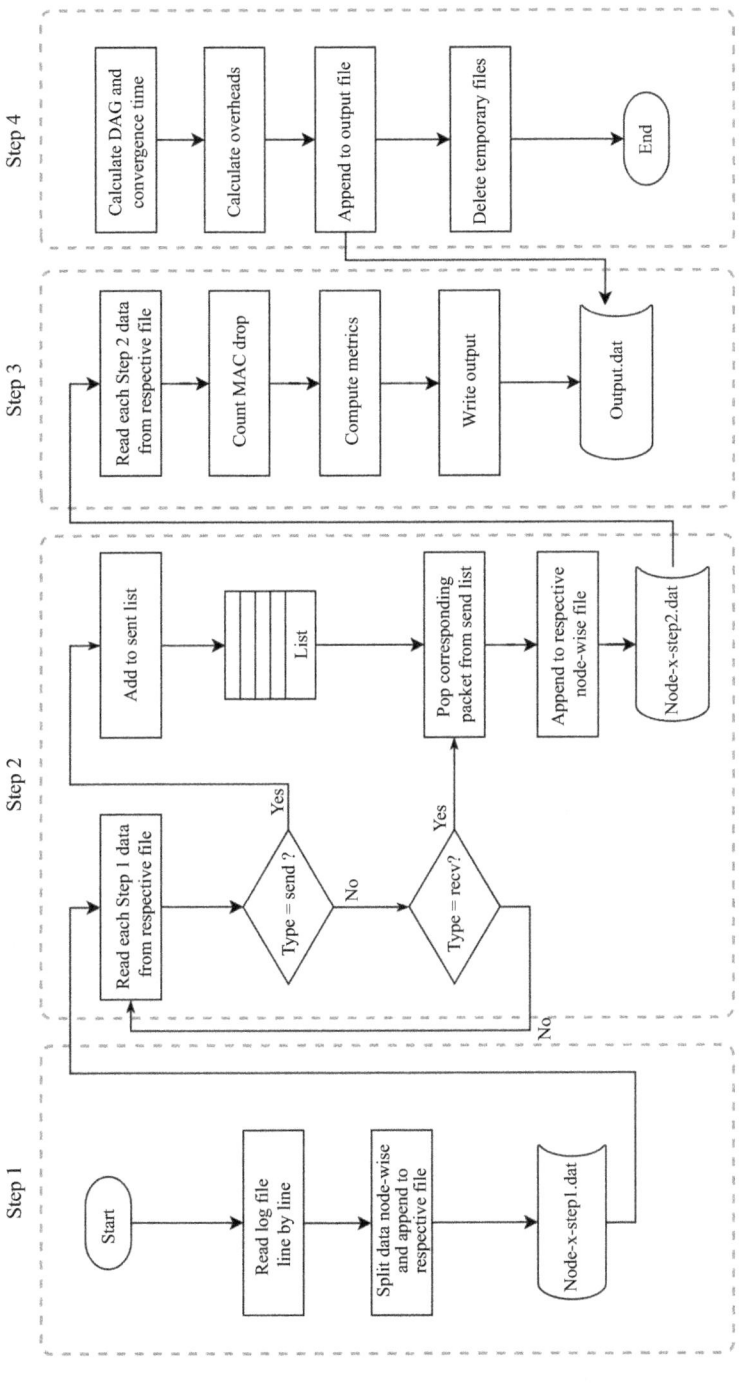

Figure 4.10   Flow chart for Algorithm 4.1

Algorithm 4.1

Input: Cooja simulation generated log file.
1.   Start
2.   Read provided data file line by line.
3.   Create data file for each node in the IoT network.
4.   Segregate each line according to node address (Node ID).
5.   Create 'Drop' file for MAC drop lines count.
6.   Repeat for all nodes
7.   Open node-wise files
8.   Search for '**send**', '**recv**' and '**drop**' in each line
9.   If '**send**' or '**recv**'
10.   Get packet ID, sent time and receive time.
11.   Create file for each node and write packet ID, send and recv time.
12.   If packet is not received, put -1 as recv time.
13.   If '**drop**'
14.   Increment drop count
15.   Write drop count in 'drop' file
16.   Open nodes file
17.   **Calculate the required mote-level and network-level metrics.**
18.   **Write results in an output file.**
19.   Delete all intermediate stage created files.
20.   End

Algorithm 4.2

Input: Cooja simulation generated log file.
1.   Start
2.   Read data file line by line until EOF
3.   Search each line for '**send**', '**drop**', overhead ('**DIS**','**DAO**','**DIO**'), '**joining DAG**'
4.   If '**send**'
5.   Increment send
6.   Check Packet ID
7.   Read Data file ahead
8.   Current Node n
9.   Find Receive packet matching ID
10.   Increment Receive
11.   **Calculate Delay, Total delay**
12.   If '**drop**'
13.   Increment MAC drop
14.   **Calculate result metrics for node n**
15.   If Overhead message with '**DIS**','**DAO**','**DIO**'
16.   Increment Overhead
17.   Current Node *n*
18.   If '**joining DAG**'
19.   Find first and last DAG time.
20.   **Find convergence time.**
21.   **Calculate Network-level metrics**
22.   Display Results
23.   End

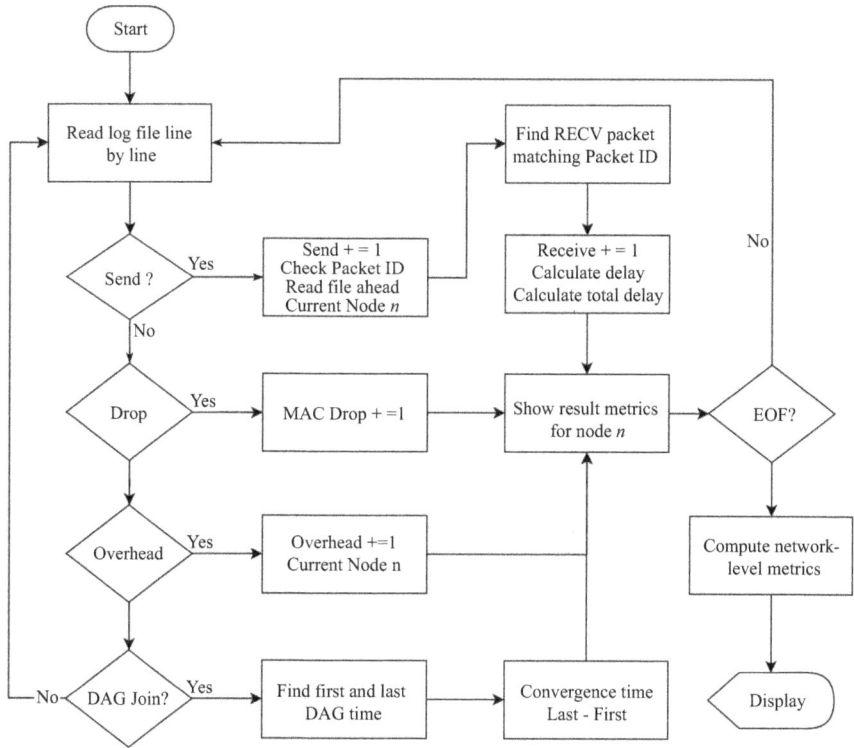

Figure 4.11 Flow chart for Algorithm 4.2

network control overhead events. Therein, it determines network convergence time by subtracting first DAG-joined event time from the last DAG-joined event time.

Algorithm 4.3 contains three classes, namely Network, Path and Packet to process the log file. A Path refers to an end-to-end connection. In Algorithm 4.3, the main logic is handled in Network class for reading file line by line and identifying packet/event type. After identifying a specific packet type, respective function is called for further processing. Regular expressions extract the information from the line of log file and handle the packet accordingly. Path class is responsible for handling end-to-end connection. It accumulates the information related to each path and dynamically adds new paths as they are detected from the log file. Paths are added dynamically and hence it can handle any number of source nodes and sinks.

Path class handles the accumulated statistics related to a single path and print-related statistics. However, network class prints the summary of all paths and network-level information, especially convergence time, network overheads and accumulated packet drops. Finally, it provides the detailed network statistics summary. Figure 4.12 shows the flow chart of Algorithm 4.3. The source log file is read exactly once and thus it completes the process in a minimum time without saving bulk information to HDD.

Algorithm 4.3

	Input: Cooja simulation generated log file.
1.	Start
2.	Declare Q, oh_pkt_stats, drop_count
3.	Check command line parameters for input file
4.	If input file is not provided open file dialog for input file
5.	Read disk file (filename)
6.	Repeat for line in lines
7.	**process packet event**
8.	time:=current line time
9.	if **sent** packet then
10.	create new packet
11.	packet.sent_time :=time
12.	add to sent Q
13.	if **recv** packet then
14.	packet := find packet in Q
15.	packet.recv_time := time
16.	packet.delay := time-packet.sent_time
17.	if **control overhead** packet then
18.	packet_type :=read packet type from line
19.	inc (oh_pkt_stats[packettype])
20.	update(dag.time)
21.	if mac **drop** packet then
22.	inc(drop_count)
23.	**Calculate overheads**
24.	delay := recv_time − sent_time
25.	jitter := avg(abs(this_delay − prev_delay))
26.	convergence_time:= last(dag.time)-first(dag.time)
27.	Display node wise metrics
28.	Display network summary
29.	Display overhead summary
30.	End

4.6   Results and discussion

Figure 4.13 depicts the network scenario which contains 1 server node running UDP-server and 10 client nodes (mote2~mote11) running UDP-client in Cooja running on Ubuntu 16.04. Table 4.3 enlists detailed simulation parameters.

It is evident from Tables 4.4–4.11 that all the proposed algorithms correctly calculate all the metrics. All three algorithms are capable to calculate detailed node-level and network-level metrics. These statistics are crucial for performance evaluation of IoT network. Three algorithms presented in this chapter can be classified according to their key characteristics. Table 4.12 presents feature comparison of proposed algorithms. Python programming language is used to develop all the algorithms. Algorithms 4.1 and 4.2 are simple scripts while Algorithm 4.3 uses Object-Oriented Programming (OOP) approach [22] for analysing the data. Algorithm 4.3 handles multipoint-to-point (MP2P), point-to-multipoint (P2MP) and point-to-point (P2P)

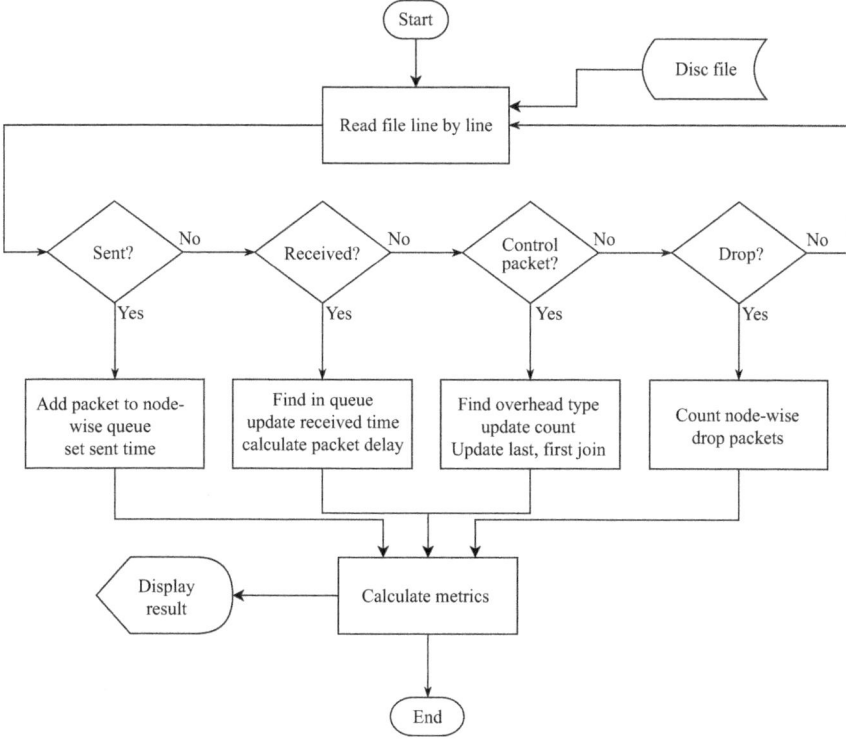

Figure 4.12   Flow chart for Algorithm 4.3

communication analysis whereas Algorithms 4.1 and 4.2 are capable of MP2P and P2P analysis. Algorithm 4.3 detects counts of motes within the data and performs the analysis accordingly in contrast to others. Algorithm 4.1 generates temporary files for intermediate stages file/data storage. Hence, it reduces the use of primary memory (RAM), and is helpful in analysing very large files with limited RAM requirements. However as a trade-off using three-fold disc storage may restrict system to fast and efficiently process the files. This approach is very effective in handling big data and dividing single data file in chunks for further processing. Algorithm 4.2 uses look ahead approach to find send and receive pairs. This approach involves holding current position and search the received packet. The drawback of this approach is that it is time-consuming and inefficient as it involves opening of log file every time to look ahead. It also incurs processing and I/O overhead. Algorithm 4.3 computes the metrics using in memory objects, due to its object-oriented architecture. The program is highly adaptive on its line pattern detection and can handle variations in application data and length of the time, node address and other parameters. This feature helps program perform better while not missing new patterns from source data. This algorithm performs far better in execution time due to controlled usage of resources.

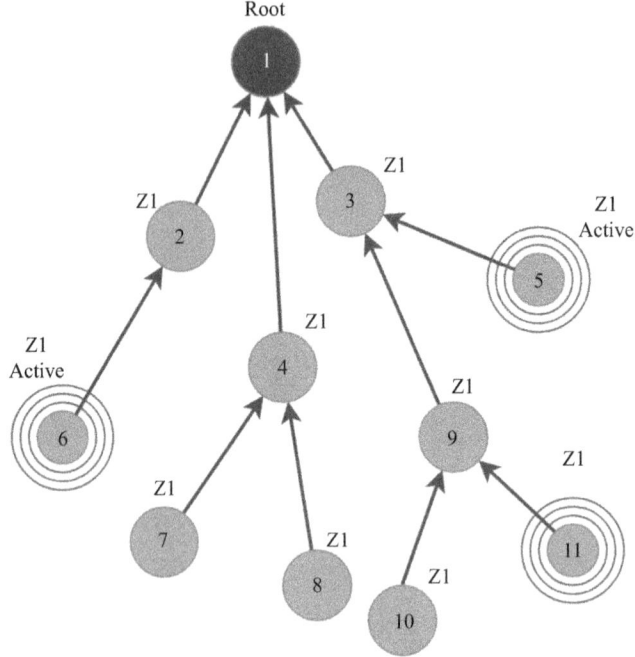

Figure 4.13   Network diagram

Table 4.3   Simulation parameters

Parameter	Value
Contiki OS version	3x
OF	MRHOF
Transmission ratio	100%
Transmission range	50 m
Mote start up delay	1.000s
Topology	Random
Simulation time	1000 sec
Client nodes	10

Since it uses primary memory to store its objects, programs still have enough margin to analyse fairly large files generated from Cooja. Algorithm 4.3 runtime is the fastest and it takes 0.095 seconds to process the log file, whereas Algorithms 4.1 and 4.2 take 0.295 seconds and 100 seconds, respectively. Algorithm 4.3 offers command line execution with parameterized source file. However. Algorithm 4.3 has a largest programming code size with 454 lines of code, whereas Algorithms 4.1 and 4.2 have 304 and 206 lines of code, respectively.

Table 4.4   Algorithm 4.1 results

Node address	Sent Packets	Received Packets	Lost Packets	Lost %	Sent bytes	Received Bytes	Lost Bytes	Lost Bytes %	Delay	Average Delay	Jitter	Frame Drop
2	307	196	111	36.16	9,210	5,880	3,330	36.16	10,739	55	28.03	–
3	291	209	82	28.18	8,730	6,270	2,460	28.18	13,107	63	30.78	52
4	305	223	82	26.89	9,150	6,690	2,460	26.89	11,732	53	28.11	–
5	289	140	149	51.56	8,670	4,200	4,470	51.56	25,307	181	80.01	137
6	304	67	237	77.96	9,120	2,010	7,110	77.96	15,235	227	72.84	1
7	305	76	229	75.08	9,150	2,280	6,870	75.08	19,907	262	87.61	2
8	307	35	272	88.60	9,210	1,050	8,160	88.60	9,384	268	107.14	–
9	303	30	273	90.10	9,090	900	8,190	90.10	8,664	289	89.10	–
10	304	33	271	89.14	9,120	990	8,130	89.14	11,430	346	129.06	–
11	307	24	283	92.18	9,210	720	8,490	92.18	6,432	268	78.00	–

Table 4.5   Algorithm 4.1 network results

Metric	Value
Sent	
Packets	3,022
Bytes	90,660
Received	
Packets	1,033
Bytes	30,990
Received %	34.2%
Lost	
Packets	1,989
Byte	59,670
Lost %	65.8%

Table 4.6   Algorithm 4.1 network overhead results

Metric	Value
Jitter %	73.068 ms
DAG	10
First DAG	5,160 ms
Last DAG	13,860 ms
Convergence Time	8,700 ms
DIS Overhead	14
DIO Overhead	136
DAO Overhead	70
Control Traffic Overhead	220
MAC Frame Dropped	192

Table 4.7 Algorithm 4.2 results

Node address	Sent Packets	Received (Packets)	Lost (Packets)	Lost % (Packets)	Sent (Bytes)	Received (Bytes)	Lost (Bytes)	Lost % (Bytes)	Delay (ms)	Average Delay (ms)	Jitter (ms)	MAC Frame drop (Packets)
2	307	196	111	36.16	9,210	5,880	3,330	36.16	10,739	55	28.03	–
3	291	209	82	28.18	8,730	6,270	2,460	28.18	13,107	63	30.78	52
4	305	223	82	26.89	9,150	6,690	2,460	26.89	11,732	53	28.11	–
5	289	140	149	51.56	8,670	4,200	4,470	51.56	25,307	181	80.01	137
6	304	67	237	77.96	9,120	2,010	7,110	77.96	15,235	227	72.84	1
7	305	76	229	75.08	9,150	2,280	6,870	75.08	19,907	262	87.61	2
8	307	35	272	88.60	9,210	1,050	8,160	88.60	9,384	268	107.14	–
9	303	30	273	90.10	9,090	900	8,190	90.10	8,664	289	89.10	–
10	304	33	271	89.14	9,120	990	8,130	89.14	11,430	346	129.06	–
11	307	24	283	92.18	9,210	720	8,490	92.18	6,432	268	78.00	–

Table 4.8 Algorithm 4.2 network results

Metric	Value
Sent	
Packets	3,022
Bytes	90,660
Received	
Packets	1,033
Bytes	30,990
Received %	34.2 %
Lost	
Packets	1,989
Byte	59,670
Lost %	65.8 %

Table 4.9 Algorithm 4.2 network overhead results

Metric	Value
Jitter %	73.068 ms
DAG	10
First DAG	5,160 ms
Last DAG	13,860 ms
Convergence Time	8,700 ms
DIS Overhead	14
DIO Overhead	136
DAO Overhead	70
Control Traffic Overhead	220
MAC Frame Dropped	192

Table 4.10   Algorithm 4.3 results

Node address	Sent (Packets)	Received (Packets)	Received %	Lost (Packets)	Lost %	Received (Bytes)	Delay (ms)	Average Delay (ms)	Jitter (ms)	MAC Drop (Packets)	Drop %
2	307	196	63.84	111	36	5,880	10,739	54	28	0	0
3	291	209	72	82	28	6,270	13,107	62	31	52	17.87
4	305	223	73.36	82	27	6,690	11,732	52	28	0	0
5	289	140	48.61	149	51	4,200	25,307	180	80	137	47.4
6	304	67	22.11	237	78	2,010	15,235	227	73	1	0.33
7	305	76	25.17	229	74	2,280	19,907	261	88	2	0.66
8	307	35	11.71	272	88	1,050	9,384	268	107	0	0
9	303	30	10.56	273	89	900	8,664	288	89	0	0
10	304	33	10.89	271	89	990	11,430	346	129	0	0
11	307	24	8.45	283	92	720	6,432	268	78	0	0
Summary	3,022	1,033	34.18	1,989	65	30,990	131,937	201	73.07	192	6.35

Table 4.11   Algorithm 4.3 network overhead results

Metric	Value		
DIS	14	Overhead packet count	220
DIO	136	First DAG time (a)	5,160 ms
DAO	70	Last DAG time (b)	13,860 ms
DAG	10	Convergence time b-a	8,700 ms
OH + data packet	3,242	Overhead %	6.79%

Table 4.12   Feature comparison of algorithms

Nomenclature	Algorithm 4.1	Algorithm 4.2	Algorithm 4.3
Metrics count	30	30	30
Node-wise stats	Yes	Yes	Yes
Network stats	Yes	Yes	Yes
Adaptive	Medium	None	High
Multi-servers	Fixed	No	Yes
Handle dynamic clients	Yes (arg)	None	Auto
Handle multiple paths	No	No	Yes
Secondary memory usage	High	Yes	No
Primary memory usage	Low	Yes	Yes
Execution time (seconds)	0.295	>100	0.095
Program architecture	script	script	Object oriented
Programming language	Python 3	Python 3	Python 3
Program distribution	File	File	Package
Program line of code	304	206	454
Need disk write permission	Yes	Yes	No
Command line execution	No	No	Yes
Command line help	No	No	Yes

4.7 Conclusion

Big data handling definitely needs specific tools for data analysis and decision-making. Data are not useful until interpreted in accurate manner. Contiki-OS simulator Cooja generates log files of IoT network simulation. All the simulation-related messages are appended in a log file as the simulation progress.

In this chapter, we proposed and compared various algorithms for Contiki-OS data analysis. Experiments show that all the proposed algorithms accurately calculate the overall metrics. However, adopted techniques for data analysis can make huge difference. All three discussed algorithms are helpful in different IoT deployment scenarios. Therefore, researchers can choose according to their preference of memory requirements, file sizes and time constraints.

References

[1] 'Internet of Things', [Online]. Available: https://data-flair.training/blogs/iot-tutorial. [Accessed 18 April 2019].

[2] 'Internet of Things', [Online]. Available: https://en.wikipedia.org/wiki/Internet_of_things. [Accessed 18 April 2019].

[3] Y. Bin Zikria, S. W. Kim, O. Hahm, M. K. Afzal, and M. Y. Aalsalem, 'Internet of Things (IoT) Operating Systems Management: Opportunities, Challenges, and Solution,' Sensors (Switzerland), vol. 19, no. 8. p. 1793, 2019.

[4] Y. B. Zikria, H. Yu, M. K. Afzal, M. H. Rehmani and O. Hahm, 'Internet of Things (IoT): Operating System, Applications and Protocols Design, and Validation Techniques', *Future Generation Computer Systems,* vol. 88, pp. 699–706, 2018.

[5] F. Al-Turjman, '5G-enabled Devices and Smart-Spaces in Social-IoT: An Overview', *Elsevier Future Generation Computer Systems,* vol. 92, no. 1, pp. 732–744, 2019.

[6] F. Al-Turjman and M. Abujubbeh, 'IoT-enabled Smart Grid via SM: An Overview', *Elsevier Future Generation Computer Systems,* vol. 96, no. 1, pp. 579–590, 2019.

[7] F. Al-Turjman and A. Malekoo, 'Smart Parking in IoT-enabled Cities: A Survey', *Sustainable Cities and Societies.* Elsevier Ltd, vol. 49, pp. 1–17, 2019.

[8] F. Al-Turjman, H. Zahmatkesh, and R. Shahroze, 'An Overview of Security and Privacy in Smart Cities' IoT Communications', *Transactions on Emerging Telecommunications Technologies*, May, pp. 1–19, 2019.

[9] F. Al-Turjman, 'Mobile Couriers' Selection for the Smart-grid in Smart Cities' Pervasive Sensing', *Elsevier Future Generation Computer Systems,* vol. 82, no. 1, pp. 327–341, 2018.

[10] A. Musaddiq, Y. B. Zikria, O. Hahm, H. Yu, A. K. Bashir and S. W. Kim, 'A Survey on Resource Management in IoT Operating Systems', *IEEE Access,* vol. 6, pp. 8459–8482, 2018. doi: 10.1109/ACCESS.2018.2808324.

[11] 'Big Data Analytics', [Online]. Available: https://www.techopedia.com/definition/27745/big-data. [Accessed 18 April 2019].

[12] 'Big Data', [Online]. Available: https://en.wikipedia.org/wiki/Big_data. [Accessed 18 April 2019].

[13] 'Contiki Operating System', [Online]. Available: https://en.wikipedia.org/wiki/Contiki. [Accessed 18 April 2019].

[14] 'Contiki: The Open Source OS for the Internet of Things', [Online]. Available: http://contiki-os.org/index.html. [Accessed 18 April 2019].

[15] 'RFC-6550 (RPL: IPv6 Routing Protocol for Low-Power and Lossy Networks)', [Online]. Available: https://tools.ietf.org/html/rfc6550. [Accessed 20 April 2019].

[16] Adam Dunkels, 'Rime—A Lightweight Layered Communication Stack for Sensor Networks', in *Wireless Sensor Networks (EWSN)*, 2007.

[17] 'Get Started with Contiki', [Online]. Available: http://www.contiki-os.org/start.html . [Accessed 18 April 2019].

[18] 'Get Started with Contiki; Step 2 Start Cooja', [Online]. Available: http://contiki-os.org/start.html. [Accessed 18 April 2019].

[19] 'RFC-6551 (Routing Metrics Used for Path Calculation in RPL)', [Online]. Available: https://tools.ietf.org/html/rfc6551. [Accessed 20 April 2019].

[20] 'RFC-6552 (Objective Function Zero for RPL)', [Online]. Available: https://tools.ietf.org/html/rfc6552. [Accessed 20 April 2019].

[21] 'The Minimum Rank with Hysteresis Objective Function', Internet Engineering Task Force (IETF) , [Online]. Available: https://tools.ietf.org/html/rfc6719. [Accessed 23 April 2019].

[22] 'Object-oriented Programming', [Online]. Available: https://en.wikipedia.org/wiki/Object-oriented_programming. [Accessed 18 April 2019].

Chapter 5

Analysis of the safety of the Internet of Things in the mesh

Mikołaj Leszczuk[1]

The Internet of Things (IoT) is an idea according to which uniquely identifiable gadgets can directly or indirectly assemble, process or transfer data via the KONNEX quick electrical installation (KNX) or a computer network. With the advent of IoT, there were also fears that IoT is growing too fast, without due consideration of the significant security challenges and regulatory changes that may be necessary. This chapter analyses the security aspects of the mesh IoT. After entering the IoT mesh theme, a general review of IoT network security is presented. Next, security aspects in the IEEE 802.15.4 standard were analysed. Significant attention was also paid to the technical guidelines that enable secure transmission of information between selected IoT mesh points. The safety of implementing the mesh IoT network has been analysed in detail. Finally, the security aspects of different systems were compared.

5.1 Introduction

The IoT is an idea according to which uniquely identifiable gadgets can directly or indirectly assemble, process, or transfer data via the KNX or a computer network (Figure 5.1*). These types of items include, among others, household appliances, lighting and heating articles, and worn (wearable) devices [1–6].

   With the advent of the IoT, there have also been concerns that the IoT is growing too fast, without taking seriously into account the security challenges [7] and the regulatory changes that may be needed [8,9]. On the IoT, most of the problems related to technical security are similar to those known from traditional servers, workstations and smartphones, but there are security challenges that only concern the IoT, including industrial security controls, hybrid systems, business processes related to IoT and end nodes [10].

   Security is the biggest problem when implementing IoT [11]. Because the IoT is spreading widely, cyberattacks, in particular, can become more and more physical

[1]AGH University of Science and Technology, Department of Telecommunications, Kraków, Poland

Figure 5.1   Drawing showing the idea of the IoT

(and not just virtual) threats [12]. IoT devices, taken over by hackers, can be used to gain unauthorized information about their users. The problem concerns devices such as televisions, kitchen appliances, cameras [13] and thermostats [14]. It has also been shown that computer-controlled devices in cars, such as brakes, engine, locks, boot lid and trunk, horn, heating and instrument panel, are vulnerable to hacker attacks that gain access to the onboard network.

In some cases, the vehicle's computer systems are connected to the Internet, allowing their remote use [15]. Until 2008, security specialists demonstrated the ability to control pacemakers without authorization remotely. Later, the hackers also showed the possibility of remote control over insulin pumps and implantable cardioverter defibrillators [16].

The US National Intelligence Council confirmed that unauthorized access to some sensors and remotely controlled facilities were obtained by foreign intelligence agents, criminals and malware makers. This means that an open market for aggregated sensor data can, of course, serve the interests of commerce and security, but equally helps criminals and spies to identify sensitive targets. In the case of some sensors, the problem of unauthorized access may be mass scale [17]. The literature of the subject emphasizes that the IoT can be a rich source of data for intelligence services [18].

Issues related to privacy and security are one of the barriers to the broader development of the IoT. The potential of the IoT for a severe invasion of privacy is an undeniable problem [19] and research on the problems of the Internet security of things is indeed not finished [19]. Among the proposed solutions in terms of the techniques used and the extent to which they meet the basic principles of privacy protection, only a few have been entirely satisfactory [19]. The industry is not very interested in security issues. Despite loud and alarming burglaries, device manufacturers remain unmoved, focusing on profitability, not on security. Meanwhile, users expect more significant control over the collected data, including the possibility of removing them, if they so choose. Without ensuring privacy, there is no wide client adoption [20].

5.2   Security of the IoT

In the IoT world, several consortia such as AllJoyn, Thread, Open Interconnect Consortium (OIC) or Industrial Internet Consortium (IIC) develop (partly competing) IoT standards. The communication/transport layer also has various measures, such as ISA100.11a, IEEE 802.15.4, NFC, ANT, Bluetooth, Eddy-stone, Zigbee, En-Ocean or WiMax. All of these standards offer different levels and schemes for implementing security. Common IoT security standards – which are also used in the broader Internet – are X.509 and Open Trust Protocol (OTrP), the former being the most common standard for managing a public key infrastructure (PKI) using digital certificates and public key encryption [21].

In this section, the two selected basic Internet security protocols will be discussed: X.509 (Section 5.2.1) and OTrP (Section 5.2.2).

5.2.1   X.509

X.509 is a standard for PKI to manage digital certificates and public key encryption. The critical part of the Transport Layer Security [22] protocol is used to secure Internet communication and e-mail.

X.509 is a standard defining a schema for public key certificates, certificate revocation and attribute certificates used to build a hierarchical PKI structure. The critical element of X.509 is the certification authority, which acts as a trusted third party about the entities and users of the certificates [23].

The concept of certificates, their validity and dismissal were presented for the first time in 1978 by Loren Kohnfelder [24]. The first version of the X.509 standard was published in 1988 [23].

International Telecommunication Union – Telecommunication Standardization Sector (ITU-T) [25] edits the original X.509 standard and describes the structure of the certificate, Certificate Revocation Lists (CRLs) and certificate of the attribute. The following standardization organizations also publish standards with identical or similar content [23]:

- Internet Engineering Task Force (IETF) as part of the Public Key Infrastructure (X.509) (PKIX) working group. The current version of the IETF standard is Request for Comments (RFC) 5280 (public key certificate and CRL) [26] and RFC 5755 (attribute certificate) [27]
- ISO as ISO/International Electrotechnical Commission (IEC) 9594-8: 2005 [28]

5.2.1.1　Strengths of X.509

X.509 is a framework for services that provide generation, distribution, control and accounting for public key certificates. This system provides secure user authentication, network traffic encryption, data integrity and non-repudiation. X.509 can efficiently manage, distribute and validate public keys included in user certificates. Most modern solutions manage digital certificates well X.509 [29].

5.2.1.2　Weaknesses of X.509

X.509 is still not widely accepted, because the cost of the registration process related to the delivery of certificates on the client's side is cumbersome. Also, managing and withdrawing certificates require very complex architecture, not to mention scalability, which is associated with the additional costs of computing resources [29].

There are many publications about X.509 problems [30–32]. The weaknesses of X.509 architecture are discussed in detail in [33]. X.509 crypt weaknesses are discussed in detail in the publications [34–37].

5.2.2　Open Trust Protocol

OTrP is a protocol used 'to install, update and delete applications and to manage security configuration in a Trusted Execution Environment (TEE)' [22].

OTrP is a protocol that complies with the Trust Execution Environment Provisioning (TEEP) architecture and provides a message protocol that includes management of trusted applications (TA) on a device with the TEE [38].

The TEE concept was designed to separate the standard operating system, also known as the Rich Execution Environment, from applications sensitive to security. In the TEE ecosystem, different device manufacturers can use different TEE implementations. Different application providers or device administrators can use different Trusted Applications Manager (TAM) service providers. It requires the establishment of an interoperational TA management protocol operating in different TEEs of different devices [38].

The TEEP [39] architecture document provides design guidelines for such an interoperable protocol. In contrast, the document [38] specifies the OTrP protocol by the architecture guidelines.

OTrP defines a mutual trust message protocol between TAM and TEE and bases on security mechanisms defined by IETF, namely JSON Web Encryption (JWE), JSON Web Signature (JWS) and JSON Web Key (JWK). Other message encoding methods [38] may also be supported.

The specification defines the data fields of messages exchanged between devices and TAM. One designed the messages in anticipation of using the most popular transport methods, such as HTTPS [38].

Each binary and configuration TA data can come from one of two sources:

1. TAM provides a signed and encrypted TA binary file and any required configuration data.
2. The client application provides the TA binary file.

The [38] specification takes into account the first case in which the public key of the recipient, which must be involved in the TAM, encrypts TA binary data and configuration data.

5.2.2.1 Strengths of the OTrP

Three strong strengths characterize OTrP.

First of all, it is an open and international protocol defining procedures for the mutual trust of devices in a network environment – the protocol based on existing open and commonly used technologies with proven reliability. As the basic base system, it uses the PKI architecture and the proven concept of the Certificate Authority (CA) [40] certification authorities.

Second, due to the use of PKI architecture, OTrP creates an open certificate market that allows applications to authenticate resources on devices. The critical requirement is to have a mechanism by which certification authorities can compete and access the devices to which they send their certificates to authenticate the resources [40].

Third, thanks to an open protocol, suppliers can create solutions for clients or servers. This approach allows an extensive and dynamic market for solution developers for both clients and servers [40].

5.2.2.2 Weaknesses of the OTrP

It deserves considering that currently, the business use of the OTrP protocol is relatively small. The standard itself currently has the status of Internet-Draft (informative), according to the IETF classification. IETF working documents are valid for a maximum of six months and may be updated, replaced or superseded by different papers at any moment. It is therefore inappropriate to use these documents as reference materials or to cite them in other ways than as 'work in progress' [38].

5.3 Safety aspects in the IEEE 802.15.4 standard

The IEEE 802.15.4 specification specifies a new class of wireless radio and protocols for low-power devices, personal area networks and sensor nodes [41]. IEEE 802.15.4 is a technical standard that defines the operation of Low-Rate Wireless Personal Area Networks (LR-WPAN). It 'specifies the physical layer and media access control for LR-WPAN and is supported by the IEEE 802.15 work-group, which defined the standard in 2003' [42]. This is the basis for the Zigbee [43] specification, ISA100.11a, WirelessHART, MiWi and SNAP – each is extending the standard by developing upper layers that are not specified in IEEE 802.15.4. Alternatively, 'IEEE 802.15.4 can be used to provide compatibility with IPv6 addressing in WPANs (IPv6 over LR-WPAN, 6LoWPAN), defining higher layers such as Thread' [44].

When it comes to secure communication, the media access control (MAC) sublayer offers functionalities that can be used by higher layers to achieve the desired level of security. Higher tier processes can specify keys to perform symmetric cryptography to protect the data field and limit its availability to a group of devices or only to a point-to-point connection; these device groups can be specified in the access-control lists (ACLs). Also, MAC calculates 'freshness checks' between successive frame receipts to ensure that likely old frames or data that are no longer considered valid are not passed to higher layers.

In addition to this safe mode, there is another unprotected MAC mode that uses only ACLs as a mechanism for deciding whether to accept frames according to their (alleged) source.

It is more deserving considering the relation of the IEEE 802.15.4 standard with the X.509 standard. The IoT industry commonly (especially in systems using the IETF [26] protocols) uses X.509 certificates and CRLs. These certificates use ASN.1 encoding, which provides excellent flexibility, but also imposes some redundancy. These certificates often have several kilobytes in length, although the cryptography data they carry usually do not exceed 512 bytes, and can even be just 64 bytes.

There is also a significant interest in using standard communication protocols in low-cost, low-speed devices. For example, it is expected that Smart Grid networks use the 802.15.4 protocol, in which the maximum packet size is 103–108 bytes [45], and the data rate is often only 20 kbit/s. Although it is highly desirable to use digital certificates on these devices and to apply universal standards such as X.509 and PKIX [26], it is also desirable to avoid redundant coding that takes up memory space and bandwidth. Both of these goals can be met thanks to Compressed X.509 Format (CXF).

Finally, it is worth mentioning that the 802.15.4 specification includes some solutions and options for security. However, there are parts of the standard for which application designers and radio designers should exercise caution when deploying and using 802.15.4 devices. In particular, some of the optional features of 802.15.4 *reduce* security.

The fundamental problems are related to the management of the initialization vector (IV), key management and insufficient protection of integrity. The issues that may occur in managing IV are the situation of the same key in many ACL entries and loss of ACL status due to power outages (no power supply or low-power operation). The second class of problems results from insufficient ACL table support for many vital models, more specifically from lack of group keys, incompatibilities between Network Shared Keying and Replay Protection, as well as from incorrectly supported pairing keys. Problems that may arise due to inadequate integrity protection are un-certified encryption modes, vulnerability to denial-of-service (DoS) service-type attacks on Advanced Encryption Standard-Counter (AES-CTR), as well as the lack of consistency in confirmation packets.

Considering the above, producers of solutions based on 802.15.4 should ignore some of these extensions (optional functions 802.15.4). There are difficulties with using the secure Application Programming Interface (API). This topic is discussed in more detail in the publication [41], in which the authors explicitly recommend to change specifications to reduce the likelihood that users deploy devices with weak security configurations.

5.4   Technical guidelines for securely transferring information between selected IoT mesh points

This section contains basic technical guidelines for the secure transmission of information between selected network points (IoT mesh), and in particular terminal nodes.

The guidelines are presented in two subsections: in terms of network operation modes (Section 5.4.1) and of communication types (Section 5.4.2).

5.4.1   Networking modes

In this subsection, due to the specificity of IoT mesh, the following three modes of network operation are considered through the prism of data protection methods:

1.   Smart metering: There is a low bit rate (about 1 kbit/s) available and cyclic communication about every 1 hour – Section 5.4.1.1.
2.   Smart lighting: There is an average bit rate (about 20 kbit/s) and cyclic communication about every 10 seconds – Section 5.4.1.2.
3.   Industry 4.0: Higher bit rate is available for embedded radio solutions (about 200 kbit/s) and cyclic communication every 1 second or more often – Section 5.4.1.3.

5.4.1.1   Smart metering

Smart meters are considered as a fundamental part of smart metering infrastructure (SMI) in smart grids. A smart meter is a digital device that uses two-way communication between the client and the tool server to exchange, manage and control consumption, e.g. energy at home. However, despite all functions, the smart meter raises some safety concerns. For example, how to exchange data between parties (e.g. an intelligent counter and a tool server), preserving customer privacy. To solve these problems, authentication and critical reconciliation in SMI must provide essential parameters of security that not only maintain trust between authorized parties but also offer other security services. The paper [46] presents the concept of crucial simplified agreement (lightweight authentication and key agreement, LAKA), which ensures trust, anonymity, integrity and adequate security in the field of the intelligent power grid. The proposed scheme uses hybrid cryptography to ensure mutual trust (authentication), dynamic session key, integrity and anonymity. The authors justify the feasibility of their proposal using a test bench using a device based on 802.15.4 (i.e. a smart meter). Also, thanks to the analysis of safety and efficiency, they show that the proposed system is more effective and energy-saving compared to previous solutions [46].

A more comprehensive overview of other smart metering technologies is available in the publication [47]. The paper presents many different techniques and standards concerning the intelligent network as well as smart measurements and Advanced Metering Infrastructures (AMI).

With the implemented method of SMI mode protection, one can think about building an AMI-integrated IoT architecture. AMI systems are systems that measure, collect and analyse distribution and consumption of media and communicate with measuring devices according to a schedule or on demand. AMI is becoming an essential part of the media distribution network and enables the development of smart cities. The publication [48] proposed an integrated IoT architecture that allows the implementation of intelligent meter networks in smart cities. The publication presents the communication protocol, data format, data collection procedure and the way of constructing a vast decision-based system based on data processing. The proposed architecture includes smart meters for electricity, water and gas.

On the other hand, expert knowledge is presented by the publication [49], which is a kind of 'tutorial', when creating intelligent networks that collect and transmit sensor data. Among the many aspects of SMI discussed, there are also issues related to the security of SMI and communication in the smart network.

It is worth noting that due to the multitude of references, the following sub-subsections contain references only to selected items. Indeed, it is worth reaching for the 'classic' review items considered to be 'primer' in the IoT area, such as [50].

5.4.1.2 Smart lighting

Market analysts also note the growing share of mesh networks for smart lighting and related solutions. Mesh networks are becoming the dominant solution in homes and commercial installations of connected devices. The relatively recent introduction of the Zigbee technology by Echo Plus underlines the growing acceptance of the mesh technology market for major consumer applications. Zigbee is used in many intelligent, home lighting systems. Thanks to the links with Nest and Google, Thread, the 802.14.5 network stack, launched in 2015, gained significant popularity. The first Thread-enabled products hit the market with the launch of the Eero Wi-Fi router. The introduction of Bluetooth mesh technology in July 2017 also highlighted the growing dynamics of mesh networks in IoT [51].

Unlike the smart metering scenario, in the smart lighting scenario it's difficult to find solutions that ensure 'total security',[†] based solely on 802.15.4. However, proposals for various hybrid solutions are available. Silicon Laboratories, for example, proposes to increase the security of intelligent lighting thanks to 802.15.4 Mesh and Bluetooth multi-protocol communication technology. By adding support for Bluetooth Low Energy (LE) to the lights in the Zigbee mesh network, the configuration of the smart lighting installation can be simplified using applications on smartphones or tablets. Bluetooth can also act as a technology to increase security. Other lighting manufacturers providing multi-protocol also provide their users with node authorization solutions (connecting them to the network) working outside the baseband of the lighting network [52].

Another solution was proposed by Cooper Lighting, cooperating with the lighting industry, which presented the LumaWatt wireless network and a control platform for external LED lighting. LumaWatt is based on the IEEE 802.15.4 standard for wireless mesh networks (WMNs) with its upper layer protocols. IEEE 802.15.4 is the basis

---

[†]The term 'total security' has been intentionally enclosed in quotes, because it only means a solution for which *so far* no security holes have been found.

for the wireless network standard based on the Zigbee mesh. Zigbee complements the 802.15.4 standardization for the upper layers of the network. However, LumaWatt uses a proprietary set of layers at 802.15.4. Among other features, the implementation includes a security feature based on 128-bit encryption [53].

5.4.1.3 Industry 4.0

In the case of the Industry 4.0 scenario, there are again no independent solutions considered to be 'completely safe'. One of the most promising technologies for safety using the IEEE 802.15.4 protocol is the hybrid, wireless-wired IoT-Bus technology developed by the Fraunhofer Institute for Integrated Circuits. IoT-Bus is a communication technology based on the adaptation of the IEEE 802.15.4 technical standard for WPAN and sensor networks for use on a wired medium. Due to the use and adjustment of the IEEE 802.15.4 standard, IoT-Bus is prepared for seamless communication between carriers between wired and wireless domains. IoT-Bus can, therefore, be extended by adding wireless nodes working in the IEEE 802.15.4 standard. To ensure compatibility with IPv6 addressing networks, the 6LoWPAN adaptive layer is used. In this way, IoT-Bus is prepared to support standard protocols such as Open Platform Communications Unified Architecture (OPC-UA), Message Queuing Telemetry Transport (MQTT) and Constrained Application Protocol (CoAP). Using the method of accessing token-based media, IoT-Bus facilitates real-time communication [54].

A more comprehensive look at the requirements and challenges of wireless connectivity for Industry 4.0 is provided in the publication [55]. The paper answers the question of what standards can meet the critical elements of Industry 4.0 wireless communication. The three main design criteria that need to be met in case of communication between devices are low latency, long working time and reliability of the connection. The paper gives examples of wireless communication methods and compares it with the requirements, along with an indication of the requirements for wireless protocols.

In turn, in the publication [56], we find a systematic review of literature and analysis for the industrial IoT. This work aims to present a formal and objective evaluation of the literature, focusing explicitly on the subject of Industry 4.0. At the outset, the authors explain the methodology used to review the system literature. Conclusions and observations are then illustrated by analysing the collected data on research questions. Finally, the strengths and limitations of work are summarized.

5.4.2 Types of communication

In this subsection, due to the specificity of IoT mesh, the following two types of communication are considered:

1. Node authorization (attach it to the network) – Section 5.4.2.1.
2. Communication between nodes and end to end – Section 5.4.2.2.

5.4.2.1 Node authorization

In the publication [57], a framework of the 6LoWPAN network access network security is proposed, which controls access of nodes to the network based on administrative authorization and enforces security compliance with authorized nodes. The proposed framework uses the Lightweight Secure Neighbour Discovery (LSEND) protocol (for securely detecting neighbours and generating key pairs), Routing Protocol for Low-Power and Lossy Networks (RPL) (for routing datagrams) and Seluge (for the dissemination of codes compliant with security rules). Unlike other access control mechanisms, the proposed solution includes an automatic repair mechanism to allow nodes to comply with security requirements, if necessary, to gain access granted by the network.

5.4.2.2   Communication between nodes and end to end

A specific feature of the IEEE 802.15.4 standard is the small packet size – 128 bytes. The header overhead is enormous, reaching a maximum of 25 bytes (including MAC), i.e. only 103 bytes of the data field remain. The problem is particularly important in the case of IPv6. Considering the 40-byte IP header and the 8-byte UDP header, *only 54 bytes remain for real data (efficiency of 42%).*

If we additionally expect to secure the data link layer, header overheads can increase by up to an additional 21 bytes. *Only 33 bytes remain for real data (26% efficiency).* This is clearly below the requirements for efficient transmission.

The solution to the problem of secure IPv6 transmission in the 802.15.4 standard is *6LoWPAN* (RFC 4944 and RFC 6282). Thanks to 6LoWPAN, packet fragmentation follows *below* the network layer. In addition, headers are also compressed: IP addresses are compressed if they can be obtained from other headers, such as the MAC header 802.15.4 (followed by prefixing for the Link-Local – fe80::, addresses are obliterated, when they can be fully deduced from the link layer addresses) and one compresses common headers (TCP, UDP, ICMP).

In the case of mesh networks, however, it must be remembered that 6LoWPAN has a '*Mesh Address Header*' that supports packet routing in a mesh network, but leaves routing details to *data link layer*. Meanwhile, 802.15.4 assumes, for a change, that mesh routing takes place in *network layer*.

Of a functional point of view, 'it is indeed interesting to look at the OpenWSN project, which aims to build a standards-based, open implementation of a complete, limited stack of network protocols for wireless sensor networks and the IoT' [58]. A more detailed discussion on the use of IEEE 802.15.4 security procedures in the OpenWSN protocol stack is included in the publication [59].

5.5   Security of the implementation of the mesh IoT

Mesh network (mesh or simply mesh-net, Figure 5.2‡) is a local network topology in which infrastructure nodes (i.e. bridges, switches and other infrastructure devices) connect directly, dynamically and not hierarchically with as many other nodes as possible and work together to route data from/to customers efficiently. This lack of

---

‡By Original: FoobazSVG: Hazmat2 – This file was derived from NetworkTopology-mesh.png, Public Domain, https://commons.wikimedia.org/w/index.php?curid=31012050

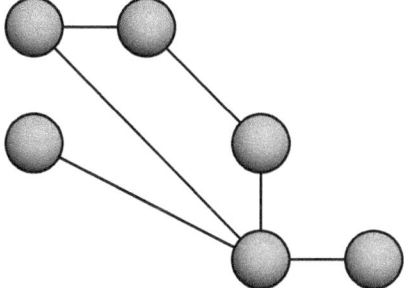

Figure 5.2   Illustration of a partial mesh network. A full mesh network would be in a situation where each node is connected to any other node in the network

dependence on one node allows each node to participate in the transmission of information. Mesh networks dynamically self-organize and configure themselves, which can reduce their installation overhead. The possibility of independent configuration allows the dynamic distribution of loads, especially in the event of failure of several nodes. This, in turn, contributes to fault tolerance and reduces maintenance costs [60].

WMN (Figure 5.3[§]) is a communication network composed of radio nodes organized in a mesh topology. It is also a form of wireless ad-hoc network [61].

WMN can be seen as a group of nodes (clients or routers) cooperating to provide connectivity. Such an open architecture, in which clients serve as routers to transfer data packets, is exposed to many types of attacks, including those that can break the entire network and cause DoS or distributed-DoS (DDoS) [62].

The following security analysis of the mesh IoT will be carried out with particular attention to the Bluetooth mesh networking protocol specifications (Section 5.5.1), Disruption-tolerant mesh network (Section 5.5.2), ELIoT Pro (Section 5.5.3), LoRa (Section 5.5.4), Thread (Section 5.5.5), Z-Wave (Section 5.5.6) and Zigbee (Section 5.5.7).

5.5.1   Bluetooth mesh networking

Wireless network with Bluetooth mesh networking (Figure 5.4[∥]), developed in 2015 [63], adopted on 13 July 2017, is a stack of Bluetooth low-energy technology-based protocols, which allows communication between multiple devices via Bluetooth [64].

A wireless network with Bluetooth mesh networking is defined in the Mesh Profile Specification [65] and the Mesh Model Specification [66].

[§]By David Johnson, Karel Matthee, Dare Sokoy, Lawrence Mboweni, Ajay Makan and Henk Kotze (Wireless Africa, Meraka Institute, South Africa) – Building a Rural Wireless Mesh Network: A Freifunk-based mesh network, CC BY-SA 2.5, https://commons.wikimedia.org/w/index.php?curid=30799980
[∥]By Bluetooth Special Interest Group – SVG rendering is drawn by me, = Nichalp Talk =, Public Domain, https://commons.wikimedia.org/w/index.php?curid=40232547

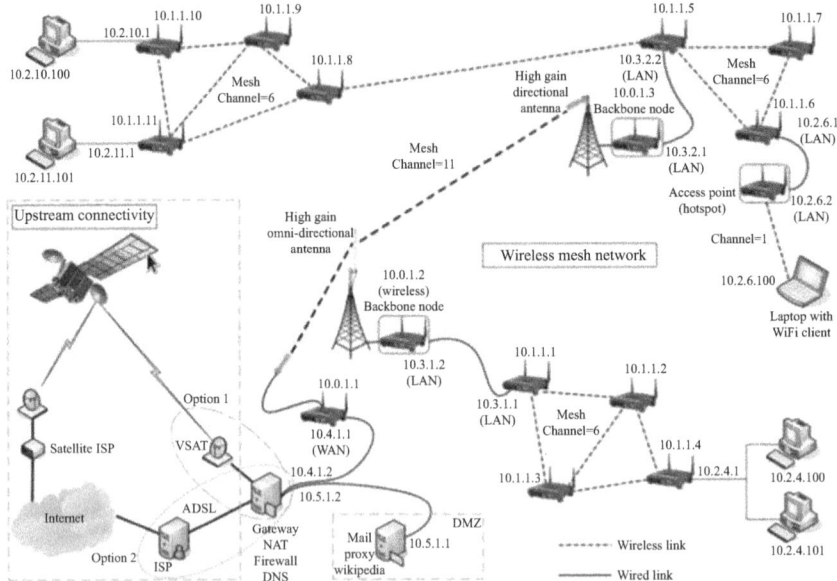

Figure 5.3   Diagram showing the possible configuration of a wireless mesh

Figure 5.4   Bluetooth Special Interest Group logo

Communication takes place in messages that can be up to 384 bytes in length, using the segmentation and reassembly (SAR) mechanism, but most signals fall into one segment, i.e. in a part of range 11 bytes. Each message starts with an operation code, which can be 1 byte (for individual messages), 2 bytes (for standard messages) or 3 bytes (for messages specific to the provider).

Each message has a source and destination address, specifying which devices process signals. Devices publish messages to destinations that can be single items/groups of things/everything else.

Each message has a sequential number that protects the network from replay attacks.

Each message is encrypted and authenticated. Two keys are used to secure the message:

1.   network key – assigned to a single mesh network;
2.   application key – specific for the given application, e.g. switching on the light or re-configuring the light.

The standard wireless network with Bluetooth mesh networking provides for the so-called 'sharing' (provisioning). Sharing is the process of installing the device on the system. This is a mandatory step to build a Bluetooth mesh network.

In the sharing process, the security component safely distributes the network key and the unique address space for the device. The sharing protocol uses the Diffie–Hellman P256 Elliptic Curve Key Exchange protocol to create a temporary key for encrypting the network key and other information. This protects against passive eavesdropping. It also provides various authentication mechanisms to protect the network information from active eavesdropping, which would use the 'man in the middle' attack during the sharing process.

The device-unique key, called the 'Device Key', comes from a secret elliptic key co-used by the provisioning device and the device being installed during the sharing process. The device key mentioned is used by the initiating module to encrypt messages for this particular device.

Sharing can be done using the Bluetooth Generic Attribute Profile (GATT) or broadcast using the specified media [65].

5.5.2 Disruption-tolerant mesh network

Disruption-tolerant mesh network, modelled and tested by Space and Naval Warfare Systems Command (SPAWAR, a branch of the US Navy) is a scale-able and secure network designed to protect strategic military resources, both stationary and mobile. In this network, control applications that operate on mesh-type nodes 'take over' when Internet connectivity is lost. Use cases include the IoT, e.g. smart drone swarms [60].

5.5.3 ELIoT Pro

ELIoT Pro (Easy & Lightweight IoT Protector), created by Cyberus Labs, is a comprehensive new generation solution in the growing market of cybersecurity of the IoT, which has the opportunity to become an industry standard in IoT. ELIoT Pro is based on the concept of an existing password-free user authentication system – Cyberus Key – an easy-to-use and secure, multi-level platform for authentication and confirmation of operations on the network, ensuring high security for online banking, e-commerce and other web and mobile applications and services. Cyberus Key was first presented in February 2017 at the Mobile World Congress in Barcelona. Cyberus Key is already available on the market and was tested and implemented by clients.

ELIoT Pro is a universal, easy-to-use authentication system designed for IoT systems, eliminating one of the weakest links in cybersecurity that are passwords and access data in human-to-machine (H2M) and machine-to-machine (M2M) communication, thus ensuring user, data and device security.

ELIoT Pro enables secure authentication and communication between devices with very little computing power, memory and the possibility of energy consumption, and therefore outside the safety frame. ELIoT Pro provides authentication and secure data transfer using the encryption algorithm developed by Cyberus Labs.

Today's IoT security systems, in particular encryption algorithms, discriminate against the vast majority of end devices (e.g. simple meters, sensors) with limited memory, computing power and battery-powered systems, ensuring safety only to those that meet the requirements of prodigious memory, significant power requirements computation and access to energy that allows one to perform advanced calculations.

ELIoT Pro provides secure data transmission in IoT networks, in areas such as user authentication in H2M communication and the case of smart network devices, also a device-to-device authentication in M2M communication to authorize commands and data flow through network devices. Using audio connection for user authentication, ELIoT Pro is an excellent system of secure authentication in systems using 'smart speakers' (so-called *Voice Activated Networks*), such as, for example, Alexa from Amazon. ELIoT Pro already today offers what is currently postulated using quantum microchips with the difference that it is a universal solution, much cheaper and easier to implement than just quantum microprocessors.

Insufficient computing power and mass storage of most devices/sensors connected to the IoT is one of the main problems/challenges related to IoT protection because these devices limit or exclude the use of current encryption methods. Computational operations and memory usage also affect battery life. Examples include sensors in remote locations, such as heat sensors or vibration sensors that are battery operated and have limited memory and computing capabilities. This leads to susceptibility to the attack of these simplest devices, and also of the entire IoT environment. ELIoT Pro provides an equally high level of security for all IoT devices, regardless of their computing/memory limits. ELIoT Pro is an agnostic system when it comes to the communication protocol used between devices and the type of IoT platform on which the system is installed. ELIoT Pro introduces an entirely new authentication and encryption algorithm that can be handled even by the simplest terminal devices. This is a technology developed by Cyberus Labs and is called ELIoT Pro's Lightweight Encryption (**ELE**, Figure 5.5¶).

5.5.4   LoRa

LoRa (*Long Ra*nge, Figure 5.6**) is a patented digital wireless data technology developed by Cycleo of Grenoble in France and acquired by Semtech in 2012 [67]. LoRa uses free, sub-gigahertz bands radio frequencies such as 169, 433, 868 (Europe) and 915 MHz (North America). LoRa enables transmissions with an extensive coverage (over 10 km in rural areas) with low-energy consumption [68]. The technology consists of two parts – LoRa, physical layer and LoRaWAN (Long-Range Wide-Area Network), upper layers.

The fundamental security aspects are ensured at the LoRaWAN layer. LoRaWAN defines a communication protocol and system architecture for the network, while the LoRa physical layer allows a long-range communication connection. The LoRaWAN layer is also responsible for managing communication frequencies, baud rates and power for all devices [69]. Devices in the network are asynchronous and transmit when they have available data to send. The data posted by the terminal device are received by

¶With permission and courtesy of Cyberus Labs
**CC0, https://en.wikipedia.org/w/index.php?curid=59270917

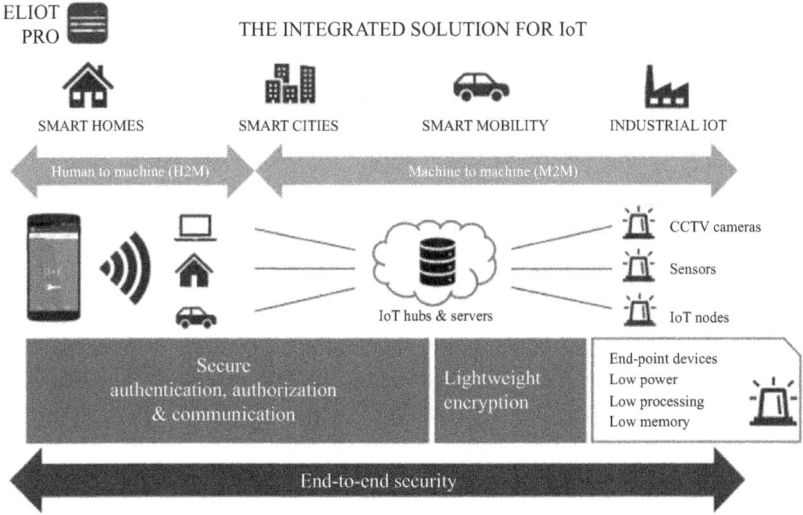

Figure 5.5   ELE – end-to-end security for IoT ecosystems

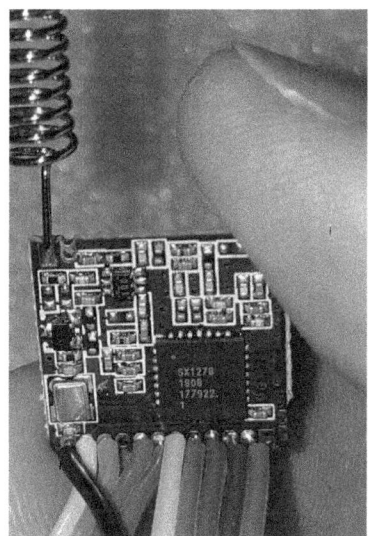

Figure 5.6   LoRa module with attached antenna and cables

multiple gateways that send data packets to the centralized network server [70]. The network server filters duplicate packages perform security checks and manage the network. The data are then transferred to the application servers [71]. The technology 'shows high reliability for the moderate load; however, it has some performance problems related to sending confirmations' [72].

5.5.5   Thread

Thread is based on IPv6 low-power network technology for IoT products to be safe and forward [73]. The Thread protocol specification is available free of charge, but this requires consent to the End-User License Agreement (EULA), which states that 'Membership in Thread Group is necessary to implement, execute and send Thread technology and Thread Group specifications' [74]. Membership in the 'Thread Group' association is subject to an annual membership fee, except the 'academics' [75].

The Thread Group announced its formation in July 2014. Thread Group is a working group of Nest Labs (Alphabet/Google affiliate), Samsung, ARM Holdings, Qualcomm, NXP Semiconductors/Free-scale, Silicon Labs, Big Ass Solutions, Somfy, OSRAM, Tyco International and Yale (a company that produces locks). Thread Group is working towards Thread becoming an industry standard that provides Thread certification for products [76]. In August 2018, Apple joined the group, hoping that this contributes to the popularization of the protocol [77].

As mentioned earlier, Thread uses the 6LoWPAN [44], which in turn uses the IEEE 802.15.4 [42] wireless protocol, providing mesh communication, much like Zigbee and other systems (Section 5.5.7). Thread, however, is addressable with IP addresses, with access to the cloud and AES encryption. Nest Labs has also released a licensed BSD license to implement Thread with an open source code (known as 'OpenThread') [78].

As mentioned above, Thread uses 6LoWPAN, which is based on the use of a connecting router, called an edge router (Thread calls its edge routers 'Border Routers'). Unlike other proprietary networks, 6LoWPAN, like any network with edge routers, does not maintain any state of the application layer because such systems send datagrams at the network layer. It means that 6LoWPAN is unaware of protocols and application changes [79]. It reduces the load on computing power for edge routers. It also means that Thread does not have to maintain the application layer. Thread Group states that many application layers can be supported as long as they are low-bandwidth and can run through IPv6 [80].

Thread Group claims that, as a rule, there is no problem of a single point of failure in the system [80]. However, if the network configuration has only one edge router, this may become a single point of failure. The edge router or another router can act as the Leader for some functions. If the Leader is damaged, another router or edge router takes its place. This is the primary way that Thread guarantees no single point of failure [80].

Thread guarantees a high level of safety. Only devices that are individually authenticated can join the network. All communication via the network is secured with the network key [80].

5.5.6   Z-Wave

Z-Wave (Figure 5.7[††]) is a wireless communication protocol primarily used for house computerization. It is 'a mesh network using low-energy radio waves for

---

[††] By Sigma Designs – http://www.z-wavealliance.org/, Public Domain, https://commons.wikimedia.org/w/index.php?curid=34336209

Figure 5.7   Logo of Z-Wave technology

communication between devices' [81], allowing wireless control of home appliances and other devices, such as lighting control, security systems, thermostats, windows, locks, swimming pools and garage door openers [82,83]. Like other protocols and systems aimed at the home and office automation market, 'the Z-Wave automation system can be controlled from a wireless remote control, wall keyboards or smartphones, tablets or computers, from a Z-Wave gateway or a central control device, acting as both a concentrator controller as well as the external portal' [83,84]. Z-Wave ensures interoperability between home manufacturers' home control systems that are part of this system. There are more and more interoperational Z-Wave products, over 1,700 in 2017 [85], and over 2,400 in 2018 [86].

Z-Wave is based on proprietary design and its main supplier is Sigma Designs [82]. In 2014, the second licensed source for Z-Wave (500 series) systems was Mitsumi [87]. Although there have been a number of academic and practical studies on home automation systems based on the Zigbee and X10 protocols, the research is still underway and their goal is to analyse the stack layer of Z-Wave protocols requiring the design of a device to capture radio packages and appropriate software to capture Z-Wave communication [88,89]. A security vulnerability has been discovered in AES Z-Wave encrypted security, which can be remotely used to unlock a door without knowledge of encryption keys, and due to changed keys subsequent network messages such as 'doors are open' are ignored by the established network controller. The vulnerability was not caused by the error of the Z-Wave protocol specification, but it was a mistake to implement the manufacturer of the door lock [89].

On 17 November 2016, Z-Wave Alliance announced higher security standards for devices receiving Z-Wave certification from 2 April 2017. The measure known as Security 2 (or S2) provides advanced security for smart home devices, gateways and hubs [90,91]. S2 includes encryption standards for transmission between nodes and requires the use of new pairing procedures for each device, with a unique PIN or QR codes on each machine. The new authentication layer is designed to prevent hackers from taking control of unsecured or poorly secured devices [92,93]. According to the Z-Wave Alliance, the new security standard is the most advanced security available on the market for smart devices and controllers, gateways and hubs [94].

5.5.7   Zigbee

Zigbee (Figure 5.8[‡‡]) is a specification compliant with IEEE 802.15.4 [42] for a set of high-level communication protocols used to create personal area networks with

---

[‡‡]By Autolycus – own work, CC BY-SA 3.0, https://commons.wikimedia.org/w/index.php?curid=25179566

Figure 5.8   Zigbee module

small, energy-saving digital radiotelephones, for example, home automation, data collection from medical devices and other devices low-power and low-bandwidth needs, designed for small projects requiring wireless connection. Zigbee is, therefore, a low-power, low-bandwidth and near-range (i.e., range in the personal area) ad-hoc network [95].

The technology specified in the Zigbee specification is intended to be simpler and cheaper than other WPANs, such as Bluetooth or more general, wireless networks, such as Wi-Fi. Zigbee applications include wireless light switches, home energy monitors, traffic management systems, and other consumer and industrial devices that require wireless low-range data transfer [95].

Low-energy consumption limits the transmission range to 10–100 metres in a straight line, depending on the output power and the environmental characteristics of [96]. Zigbee devices can also send data over long distances by passing data through a grid of intermediate devices to reach those more distant. Zigbee is usually used in low-bandwidth applications that require a long battery life and a secure network (Zigbee networks are secured with 128-bit symmetric encryption keys). Zigbee has a defined transmission rate of 250 kbit/s, best suited to variable data transmission from a sensor or input device [95].

Zigbee is based on the physical layer and MAC defined in the IEEE 802.15.4 [42] standard for low-frequency personal wireless networks. The specification includes four additional key components: network layer, application layer, Zigbee Device Objects (ZDO) and application objects defined by the manufacturer. ZDOs are responsible for specific tasks, including device role tracking and management of network connection requests, as well as device discovery and security [95].

As one of its characteristics, Zigbee provides devices with secure communication, protecting the creation and transport of cryptography keys, encrypting frames and controlling devices. It is based on the underlying security rules defined in IEEE 802.15.4 [42]. This part of the architecture is based on the correct management of symmetric keys and the proper implementation of security methods and policies [95].

Table 5.1   Security specifications for protocols (based on [97])

Protocol	security
BLE	Confidentiality, authentication and key derivation
LoRa	Multi-step encrypted AES
Thread	Banking-class, public-key cryptography
Wi-Fi	Encryption of protected access to Wi-Fi networks (WPA2)
Z-Wave	AES encryption
Zigbee	AES encryption, Cipher Block Chaining – Message Authentication Code (CBC-MAC)

5.6   Comparison of security aspects in different networks

This part gives a parallel study of the security of the Bluetooth Low Energy home network (BLE), Thread, Wi-Fi, Z-Wave and Zigbee protocols, with particular reference to the last two.

Table 5.1 presents the basic security specifications of protocols.

It is worth commenting here that when it comes to security, both Z-Wave and Zigbee use the same symmetric encryption AES-128 [98], standing at the same level of protection as large banks use [99], and their developers claim that these solutions are safe and impossible to break. Nothing is, of course, 100% safe, but one should know that these two high standards take the same, reliable approach [98].

The surprising advantage of the mesh network, which may be somewhat controversial, is that it is considered a safer option than a system with a concentrator. When a user has a device communicating through one dedicated concentrator, instead of devices communicating directly with each other (before they connect to the cloud), the potential vulnerabilities to attack are more significant.

Of course, this means that one attack can potentially take control of all devices. However, an attacker would probably have a tough task and would need much more resources and knowledge to do it than if a network that only provides underlying physical security would be attacked.

It is also worth mentioning that, for example, each Z-Wave network and each device in each system is assigned unique identifiers that are used when communicating with the concentrator. Thanks to this, other devices are theoretically unable to control the connected devices. In practice, however, devices such as door locks and alarms require a higher level of security [99].

Zigbee, Wi-Fi and Bluetooth usually work in the 2.4 GHz network. However, the real benefit that Zigbee has over Wi-Fi and Bluetooth is increased security. Zigbee uses network keys to ensure device security. However, standards differ because many different companies manufacture devices using Zigbee technology. Zigbee device manufacturers are criticized for using only minimal security measures, especially after it has been shown that even a door lock is easily susceptible to potential hackers [99].

5.7 Summary

This chapter analyses the security aspects of the mesh IoT. After introducing into the IoT mesh theme, a general analysis of the safety of the IoT was presented. Next, security aspects in the IEEE 802.15.4 standard were analysed. Significant attention was also paid to the technical guidelines that enable secure transmission of information between selected IoT mesh points. The security of the implementation of the mesh IoT network was analysed in detail. Finally, the security aspects of different systems were compared.

References

[1]   Wikipedia. Internet of Things – Wikipedia. Wikipedia. 2019.

[2]   Al-Turjman F. 5G-enabled devices and smart-spaces in social-IoT: An overview. Future Generation Computer Systems. 2019;92:732–744. Available from: http://www.sciencedirect.com/science/article/pii/S0167739X1731 1962.

[3]   Al-Turjman F, and Abujubbeh M. IoT-enabled smart grid via SM: An overview. Future Generation Computer Systems. 2019;96:579–590. Available from: http://www.sciencedirect.com/science/article/pii/S0167739X1831759X.

[4]   Al-Turjman F, Ever E, and Zahmatkesh H. Small cells in the forthcoming 5G/IoT: Traffic modelling and deployment overview. IEEE Communications Surveys Tutorials. 2019 Firstquarter;21(1):28–65.

[5]   Al-Turjman F, and Malekloo A. Smart parking in IoT-enabled cities: A survey. Sustainable Cities and Society. 2019;49:101608. Available from: http://www.sciencedirect.com/science/article/pii/S2210670718327173.

[6]   Al-Turjman F. The road towards plant phenotyping via WSNs: An overview. Computers and Electronics in Agriculture. 2019;161:4–13. BigData and DSS in Agriculture. Available from: http://www.sciencedirect.com/science/article/pii/S0168169917307482.

[7]   Singh J, Pasquier T, Bacon J, *et al.* Twenty security considerations for cloud-supported Internet of Things. IEEE Internet of Things Journal. 2016;3(3): 269–284.

[8]   Clearfield C. Why the FTC can't regulate the Internet of Things. Forbes. 2013.

[9]   Feamster N. Mitigating the increasing risks of an insecure Internet of Things. Freedom to Tinker. 2017.

[10] Li S. Chapter 1 – Introduction: Securing the Internet of Things. In: Li S, Xu LD, editors. Securing the Internet of Things. Boston: Syngress; 2017. pp. 1 – 25. Available from: http://www.sciencedirect.com/science/article/pii/B978012 8044582000019.

[11] Weissman C. We asked executives about The Internet of Things and their answers reveal that security remains a huge concern. Business Insider. 2015.

[12] Clearfield C. Rethinking security for the Internet of Things. Harvard Business Review. 2013.

[13]   Witkovski A, Santin A, Abreu V, *et al*. An IdM and key-based authentication method for providing single sign-on in IoT. In: 2015 IEEE Global Communications Conference (GLOBECOM); 2015. pp. 1–6.

[14]   Steinberg J. These devices may be spying on you (even in your own home). Forbes. 2014.

[15]   Greenberg A. Hackers remotely kill a jeep on the highway—With me in it. Wired. 2015.

[16]   Loukas G. 3 – Cyber-physical attacks on implants and vehicles. In: Loukas G, editor. Cyber-Physical Attacks. Boston: Butterworth-Heinemann; 2015. pp. 59–104. Available from: http://www.sciencedirect.com/science/article/pii/B9780128012901000035.

[17]   Intelligence SCB. Disruptive Technologies Global Trends 2025. Energy Storage Materials. 2010.

[18]   Ackerman S. CIA Chief: We'll spy on you through your dishwasher. Wired. 2012.

[19]   Aleisa N, and Renaud K. Privacy of the Internet of Things: A systematic literature review (extended discussion). CoRR. 2016;abs/1611.03340. Available from: http://arxiv.org/abs/1611.03340.

[20]   Basenese L. The best play on the Internet of Things trend. Wall Street Daily. 2015.

[21]   Mehnen J, He H, Tedeschi S, *et al*. Practical security aspects of the Internet of Things. In: Practical Security Aspects of the Internet of Things. Cham: Springer International Publishing; 2017. pp. 225–242. Available from: https://doi.org/10.1007/978-3-319-50660-9_9.

[22]   Postscapes. IoT standards and protocols. Postscapes. 2017.

[23]   Wikipedia. X.509 – Wikipedia. Wikipedia. 2019.

[24]   Kohnfelder LM. Towards a practical public-key cryptosystem. Massachusetts Institute of Technology; 1978.

[25]   ITU-T, editor. ITU-T X.509 : Information technology – Open systems interconnection – The Directory: Public-key and attribute certificate frameworks. International Telecommunication Union; 2010.

[26]   Boeyen S, Santesson S, Polk T, *et al*., editors. Internet X.509 Public Key Infrastructure Certificate and Certificate Revocation List (CRL) Profile. RFC Editor; 2008. RFC 5280. Available from: https://rfc-editor.org/rfc/rfc5280.txt.

[27]   Turner S, Farrell S, Housley R. Turner S, Farrell S, and Housley R, editors. An Internet Attribute Certificate Profile for Authorization. RFC Editor; 2010. RFC 5755. Available from: https://rfc-editor.org/rfc/rfc5755.txt.

[28]   ISO, editor. Information technology – Open systems interconnection – The Directory: Public-key and attribute certificate frameworks. ISO; 2005. http://www.iso.org/iso/iso_catalogue/catalogue_tc/catalogue_detail.htm?csnumber=43793.

[29]   Cobb M. The strengths and weaknesses of PKI and PGP systems. TechTarget. 2006.

[30]   Ellison C, and Schneier B. Ten risks of PKI: What you're not being told about public key infrastructure. Computer Security Journal. 2000. Available

from: https://www.schneier.com/academic/archives/2000/01/ten_risks_of_pki_wha.html

[31] Gutmann P. PKI: It's not dead, just resting. Computer. 2002;35(8):41–49.

[32] Gutmann P. Everything you never wanted to know about PKI but were forced to find out. University of Aukland, presentation. 2002.

[33] Zusman M, and Sotirov A. Sub-prime PKI: Attacking extended validation SSL. Black Hat Security Briefings. 2009.

[34] Lenstra A, and De Weger B. On the possibility of constructing meaningful hash collisions for public keys. In: Australasian Conference on Information Security and Privacy. Springer; 2005. pp. 267–279.

[35] Sotirov A, Stevens M, Appelbaum J, *et al.* MD5 considered harmful today, creating a rogue CA certificate. In: 25th Annual Chaos Communication Congress; 2008. p. 40.

[36] MCDONALD C. Automatic Differential Path Searching for SHA-1 Rump Session in Eurocrypt2009. http://eurocrypt2009rumpcrypto/. 2009.

[37] Kessler GC. An overview of cryptography (Updated Version, 3 March 2016). The Handbook on Local Area Networks, Auerbach. 2016.

[38] Pei M, Atyeo A, Cook N, *et al.* The Open Trust Protocol (OTrP). Internet Engineering Task Force; 2018. draft-ietf-teep-opentrustprotocol-02. Work in Progress. Available from: https://datatracker.ietf.org/doc/html/draft-ietf-teep-opentrustprotocol-02.

[39] Pei M, Tschofenig H, Wheeler D, *et al.* Trusted Execution Environment Provisioning (TEEP) Architecture. Internet Engineering Task Force; 2018. Work in Progress. Available from: https://datatracker.ietf.org/doc/html/draft-ietf-teep-architecture-01.

[40] Canel M. Exploring the Open Trust Protocol. Black Hat Security Briefings. 2017.

[41] Sastry N, and Wagner D. Security considerations for IEEE 802.15.4 networks. In: Proceedings of the 3rd ACM Workshop on Wireless Security. WiSe'04. New York, NY, USA: ACM; 2004. pp. 32–42. Available from: http://doi.acm.org/10.1145/1023646.1023654.

[42] IEEE. IEEE standard for information technology – Telecommunications and information exchange between systems – Local and metropolitan area networks – Specific requirements Part 15.4: Wireless Medium Access Control (MAC) and Physical Layer (PHY) Specifications for Low-Rate Wireless Personal Area Networks (WPANs). IEEE; 2006. Available from: http://dx.doi.org/10.1109/ieeestd.2006.232110.

[43] Gascón D. Security in 802.15.4 and ZigBee networks. Libelium World. 2009.

[44] Shelby Z, and Bormann C. 6LoWPAN: The Wireless Embedded Internet. Wiley Series on Communications Networking & Distributed Systems. Wiley; 2011. Available from: https://books.google.pl/books?id=3Nm7ZCxscMQC.

[45] Montenegro G, Hui J, Culler D, *et al.*, editors. Transmission of IPv6 Packets over IEEE 802.15.4 Networks. RFC Editor; 2007. RFC 4944. Available from: https://rfc-editor.org/rfc/rfc4944.txt.

[46] Kumar P, Gurtov A, Sain M, *et al.* Lightweight authentication and key agreement for smart metering in smart energy networks. IEEE Transactions on Smart Grid. 2018;p. 1.

[47] Jain S, Kumar V, Paventhan A, *et al.* Survey on smart grid technologies – Smart metering, IoT and EMS. In: Electrical, Electronics and Computer Science (SCEECS), 2014 IEEE Students' Conference on. IEEE; 2014. pp. 1–6.

[48] Lloret J, TomÃ¡s J, Canovas A, *et al.* An integrated IoT architecture for smart metering. IEEE Communications Magazine. 2016;54:50–57.

[49] Kayastha N, Niyato D, Hossain E, *et al.* Smart grid sensor data collection, communication, and networking: A tutorial. Wireless Communications and Mobile Computing. 2014;14(11):1055–1087.

[50] Xia F, Yang LT, Wang L, *et al.* Internet of Things. International Journal of Communication Systems. 2012;25(9):1101–1102.

[51] Grieshaber M. Tyler N, Ring P, and Grylls B, editors. Achieving the Vision of 'IoT in Everything'. MA Business Ltd; 2017. http://www.newelectronics.co. uk/electronics-technology/achieving-the-vision-of-iot-in-everything/166277/.

[52] Silicon Laboratories. Enhancing Smart Lighting with 802.15.4 Mesh, Bluetooth and Multiprotocol Connectivity. Silicon Laboratories. 2019.

[53] Wright M. Cooper's LED LumaWatt controls to be based on IEEE 802.15.4 mesh. LEDs Magazine. 2014.

[54] Loske M, Oeder A, and Klatt M. IoT-Bus for micro-grid control and local energy management based on the IEEE Std. 802.15.4. Electrical Engineering. 2016;98(4):363–368. Available from: https://doi.org/10.1007/s00202-016-0426-x.

[55] Varghese A, and Tandur D. Wireless requirements and challenges in Industry 4.0. In: 2014 International Conference on Contemporary Computing and Informatics (IC3I); 2014. pp. 634–638.

[56] Liao Y, Loures E de FR, and Deschamps F. Industrial Internet of Things: A systematic literature review and insights. IEEE Internet of Things Journal. 2018;1–1. Available from: https://ieeexplore.ieee.org/document/8355897.

[57] Oliveira LML, Rodrigues JJPC, de Sousa AF, *et al.* A network access control framework for 6LoWPAN networks. Sensors. 2013;13(1):1210–1230. Available from: http://www.mdpi.com/1424-8220/13/1/1210.

[58] Watteyne T, Vilajosana X, Kerkez B, *et al.* OpenWSN: A standards-based low-power wireless development environment. Trans Emerging Telecommunications Technologies. 2012;23(5):480–493. Available from: http://dblp.uni-trier.de/db/journals/ett/ett23.html#WatteyneVKCWWGP12.

[59] Sciancalepore S, Piro G, Boggia G, *et al.* Application of IEEE 802.15.4 Security Procedures in OpenWSN protocol stack. IEEE Standards Education e-Magazine. 2014 12;4.

[60] Wikipedia. Mesh networking – Wikipedia. Wikipedia. 2019.

[61] Toh CK. Ad Hoc Mobile Wireless Networks: Protocols and Systems. Englewood Cliffs, NJ 07632, USA: P T R Prentice-Hall; 2002. Available from: http://www.phptr.com/ptrbooks/ptr_0130078174.html.

[62]   Alanazi S, Saleem K, Al-Muhtadi J, *et al.* Analysis of denial of service impact on data routing in Mobile eHealth Wireless Mesh Network. Mobile Information Systems. 2016:4853924:1–4853924:19. Available from: https://new.hindawi.com/journals/misy/2016/4853924/.

[63]   Hegenderfer S. Get Ready for Bluetooth Mesh! Blog. 2016.

[64]   www bluetooth com, editor. Low Energy: Mesh | Bluetooth Technology Website. www.bluetooth.com; 2019. https://www.bluetooth.com/what-is-bluetooth-technology/how-it-works/le-mesh.

[65]   Mesh Working Group. Mesh Profile Bluetooth® Specification; 2017.

[66]   Mesh Working Group. Mesh Model Bluetooth® Specification; 2017.

[67]   Prajzler V. LoRa, LoRaWAN and LORIOT.io. LORIOTio. 2015.

[68]   Sanchez-Iborra R, Sanchez-Gomez J, Ballesta-Viñas J, *et al.* Performance evaluation of LoRa considering scenario conditions. Sensors. 2018;18(3):772.

[69]   www lora-alliance org, editor. LoRaWAN for Developers. www.lora-alliance.org; 2019. https://lora-alliance.org/lorawan-for-developers.

[70]   www link-labs com, editor. A Comprehensive Look At LPWAN For IoT Engineers & Decision Makers. www.link-labs.com; 2019. https://www.link-labs.com/lpwan.

[71]   Alliance L, editor. LoRaWAN What is it. Technical Marketing Work-group 1.0. Alliance, LoRa; 2015.

[72]   Bankov D, Khorov E, and Lyakhov A. On the limits of LoRaWAN channel access. In: Proceedings of the 2016 International Conference on Engineering and Telecommunication (EnT), Moscow, Russia; 2016. pp. 29–30.

[73]   Group T, editor. About. Thread Group; 2015. https://www.threadgroup.org/What-is-Thread/Overview.

[74]   Group T, editor. Thread 1.1 Specification. Thread Group; 2019. https://www.threadgroup.org/ThreadSpec.

[75]   Group T, editor. Thread Group. Thread Group; 2019. https://www.thread group.org/thread-group#Membershipbenefits.

[76]   Randewich N. Google's Nest launches network technology for connected home. Reuters. 2014 July.

[77]   Miller C. Apple joins 'The Thread Group', opening up the possibility of more advanced HomeKit tech. 9to5Mac. 2018 08.

[78]   Labs N, editor. OpenThread. Nest Labs; 2019. https://github.com/openthread/openthread.

[79]   Olsson J. 6LoWPAN demystified. Texas Instruments. 2014;13.

[80]   Group T, editor. Thread Stack Fundamentals. Thread Group; 2015. https://threadgroup.org/Portals/0/documents/whitepapers/Thread% 20Stack%20Fundamentals_v2_public.pdf.

[81]   Stark H. The ultimate guide to building your own smart home in 2017. Forbes. 2017 May.

[82]   Frenzel L. What's the Difference between ZigBee and Z-Wave? Electronic Design. 2012 March.

[83]   Kaven O. Zensys' Z-Wave Technology. PC Magazine. 2005 January.

[84] Sciacca J. Smarten up your dumb house with Z-Wave automation. Digital Trends. 13 11. Available from: https://www.digitaltrends.com/home/smarten-dumb-house-z-wave-automation/.

[85] Pink R. ZigBee vs Z-Wave for the IoT. Electronics 360. 2017 May.

[86] Silicon Laboratories. Silicon Labs Completes Acquisition of Sigma Designs' Z-Wave Business. Silicon Laboratories. 2018.

[87] Miller M. Z-Wave becomes multi-sourced standard as home control market heats up globally. sigmadesignscom. 2014 January.

[88] Picod JM, Lebrun A, and Demay JC. Bringing software defined radio to the penetration testing community. In: Black Hat USA; 2014. pp. 1–7.

[89] Fouladi B, and Ghanoun S. Security evaluation of the Z-Wave wireless protocol. In: Black Hat USA. vol. 24. Black Hat; 2013. pp. 1–2.

[90] Hamilton L. Z-Wave Alliance announces board member and new security mandate. CED Magazine. 2016 December.

[91] Wong W. Q&A: S2's Impact on Z-Wave and IoT security. Electronic Design. 2017 January.

[92] Crist R. Z-Wave smart-home gadgets announce new IoT security standards. CNET. 2016 November.

[93] Crist R. Your Z-Wave smart home gadgets just got more secure. CNET. 2017 April.

[94] Briodagh K. Mandatory security implementation for Z-Wave IoT devices takes effect. IoT Evolution. 2017 April.

[95] Wikipedia. Zigbee – Wikipedia. Wikipedia. 2019.

[96] Zigbee Alliance, editor. ZigBee Specification FAQ. ZigBee Alliance; 2013. https://web.archive.org/web/20130627172453/http://www.zigbee.org/Specific ations/ZigBee/FAQ.aspx.

[97] O Morales J, Lopez NA, Parado J, *et al.*. Lopez NAT, editor. A comparative study of Thread against ZigBee, Z-Wave, Bluetooth, and Wi-Fi as a home-automation networking protocol. ResearchGate; 2016.

[98] Charara S. Zigbee vs Z-Wave: Two big smart home standards explored. The Ambient. 2018.

[99] Brown S. Zigbee vs. Z-Wave Review: What's the Best Option for You? Safewise. 2018.

Chapter 6

Design of smart urban drainage systems using evolutionary decision tree model

Mir Jafar Sadegh Safari[1] and Ali Danandeh Mehr[2]

Recently, as an alternative method for monitoring of drainage systems, Internet of Things (IoT) technology is initiated in smart cities. IoT is used for detection of the location of the sediment deposition within the drainage pipe system to alert for repairing before complete blocking. However, from the hydraulic point of view, it is reasonable to design the drainage and sewer pipes to prevent the deposition of the sediment based on the physical parameters. To this end, instead of detection of blockage location, monitoring the flow characteristics is of more importance to keep pipe bottom clean from sediment deposition. Accordingly, smart sensors mounted in the drainage and sewer pipes should read the flow velocity and alert once the flow reaches a velocity in which sediment deposition is occurred. In order to determine the sediment deposition velocity, this study models sediment transport in drainage systems by means of evolutionary decision tree (EDT) technique. EDT results are compared with conventional decision tree (DT) and evolutionary genetic programming (GP) techniques. A large number of experimental data covering wide ranges of sediment and pipe size were used for the modeling. Evaluation of the developed models in terms of verity of statistical indices showed the outperformance of the proposed EDT model. The EDT, DT and GP models were found superior to their traditional corresponding regression models existing in the literature. Results are helpful for determination of the flow characteristics at sediment deposition condition in drainage systems maintained using IoT technology in smart cities.

6.1   Introduction

Sediment transport modeling in drainage and sewer systems has been the subject of studies in recent decades in the aim of the self-cleansing pipe design. Uncontrolled sediment deposition in the sewer pipes causes numerous problems such as pollution, blockage and surcharging the sediment in the flow. Modeling of sediment transport

---

[1]Department of Civil Engineering, Yasar University, Izmir, Turkey
[2]Department of Civil Engineering, Antalya Bilim University, Antalya, Turkey

in rigid boundary open channels is of importance for designing all types of rigid boundary channels. Self-cleansing design criteria can be classified based on different sediment transport modes as incipient motion, incipient deposition and nondeposition with and without deposited bed conditions. Among these, nondeposition condition is mostly considered for design of drainage systems to prevent continued deposition of sediment at the channel bottom.

According to the most drainage system standards in many countries, minimum flow velocity or minimum bed shear stress required for retention of sediment in motion within flow is used as the self-cleansing design criteria. This is the conventional design criteria where a single value of velocity or shear stress was used based on experience and without deep theoretical rationalization. Available design criteria are reported by the Construction Industry Research and Information Association in the UK [1] and Safari [2]. Minimum velocity is the most widely used design criterion for the self-cleansing channels within the range of 0.3–1 m/s. The minimum design velocities change in each country and based on type of sewer (storm, sanitary, combined). This method is mostly used in the USA and European countries such as France, Germany and UK (see [3–5]). Many important factors are missing in this method like quantity and type of sediment and sewer size. Table 6.1 gives a summary of available design criterion based on minimum self-cleansing velocity as adopted by different countries.

Similar to the minimum velocity, minimum shear stress criterion is used in some countries such as the USA, the UK, Norway, Germany and Sweden in the range of 1–12.6 N/m^2 (see [3–5]). The minimum shear stress method has deficiencies similar to minimum velocity method. Table 6.2 gives a summary of available design criterion based on minimum self-cleansing shear stress used in various countries. In both Tables 6.1 and 6.2, water (i.e., rain or snow) which lands on a surface (roof, patio, drive, roadway, etc.) should run into the surface water and transport with storm

Table 6.1   Minimum self-cleansing velocity required for nondeposition [2]

References	Country	Sewer type	Minimum velocity (m/s)
ASCE [6]	USA	Sanitary	0.6
		Storm	0.9
British Standard BS8005 [7]	UK	Storm	0.75
		Combined	1.0
Minister of Interior [8]	France	Sanitary	0.3
		Combined	0.6
		Separate	0.3
European Standard EN 752-4 [9]	Europe	All sewers	0.7 once/day for pipe $D < 300$ mm 0.7 or more if necessary for pipe $D > 300$ mm
Abwassertechnische Verreinigung ATV, Standard A 110 [10] replaced by ATV-DVWK-Regelwerk [11]	Germany	Sanitary Storm Combined	Depends on pipe diameter ranging from 0.48 ($D = 150$ mm) to 2.03 ($D = 3,000$ mm)
DID [12]	Malaysia	Storm	0.6

Table 6.2 Minimum self-cleansing shear stress required for nondeposition [2]

References	Country	Sewer type	Minimum shear stress (N/m²)
Lysne [13]	USA	...	2.0–4.0
ASCE [6]	USA	...	1.3–12.6
Yao [14]	USA	Storm	3.0–4.0
		Sanitary	1.0–2.0
CIRIA [1]	UK	...	6.2
Lindholm [15]	Norway	Combined	3.0–4.0
		separate	2.0
Scandiaconsult [16]	Sweden	All	1.0–1.5
Brombach *et al.* [17]	Germany	Combined	1.6 to transport 90% of all sediment

sewer systems. Wastewater from anything else (e.g., toilets, baths, showers, sinks and washing machines) that contains contaminants (whether that's human waste or detergents) should go into the sanitary sewer network.

Rather than just using a single value, nondeposition design concept was further modified to use more parameters in the 1990s which resulted in the nondeposition without deposited bed, nondeposition with deposited bed and incipient deposition concepts [18]. This study aims to analyze nondeposition without deposited bed criterion applicable in smart cities coupling with IoT technology. To this end, nondeposition studies with a clean bed channel are highlighted herein.

6.1.1 Overview of drainage systems design criteria

As a primary studies in the literature, Novak and Nalluri [19,20] compared sediment transport in loose and rigid boundary channels and pointed out that in loose boundary channels, higher shear stress and velocity are needed to keep sediment particles in motion. Performing experiments in rigid boundary channels, Macke [21] and Arora [22] investigated the suspended load sediment transport in which their results were analyzed then by Nalluri and Spaliviero [23]. As comprehensive studies, May *et al.* [24] and May [25] carried out experiments in pipes to develop bed load models by means of shear stress role on the surface layer of the bed considering hydrodynamic forces exerted on a particle. Ackers and White [26] theory is one of the most common methods for analyzing sediment transport in loose boundary channels. Ackers [27] considered effective bed width and made a modification on Ackers and White [26] method to apply in rigid boundary channels.

Mayerle [3] and Mayerle *et al.* [28] studied sediment transport at limit of deposition condition in circular and rectangular cross-section channels and developed

$$\frac{V}{\sqrt{gd(s-1)}} = 14.43 C_v^{0.18} D_{gr}^{-0.14} \left(\frac{d}{R}\right)^{-0.56} \lambda^{0.18} \tag{6.1}$$

as self-cleansing model applicable for drainage system design, where V is flow mean velocity; g gravitational acceleration, d sediment median size, s relative specific mass of sediment to fluid, C_v sediment volumetric concentration, R hydraulic radius, λ friction factor and D_{gr} dimensionless grain size parameter defined as:

$$D_{gr} = \left(\frac{(s-1) \, gd^3}{v^2} \right)^{1/3} \tag{6.2}$$

where v is fluid kinematic viscosity. Ab Ghani [29] investigated the effect of pipe size on nondeposition condition of sediment transport and recommended:

$$\frac{V}{\sqrt{gd(s-1)}} = 3.08 C_v^{0.21} D_{gr}^{-0.09} \left(\frac{R}{d} \right)^{0.53} \lambda^{-0.21} \tag{6.3}$$

as a bed load sediment transport model. Through evaluation of different pipe size experimental data, Ab Ghani [29] found out that channel design velocity increases while the pipe size increases. The applicability of aforesaid models was evaluated by the number of researchers in practice [30–32]. Ota [33] studied the effect of sediment particle characteristics at nondeposition sediment transport and recommend practical models [34]. Vongvisessomjai *et al.* [5] simplified formerly developed self-cleansing models and suggested:

$$\frac{V}{\sqrt{gd(s-1)}} = 4.31 C_v^{0.226} \left(\frac{d}{R} \right)^{-0.616} \tag{6.4}$$

for bed load sediment transport by conducting experiment in two circular pipes. Safari *et al.* [35] focused on the effect of channel cross-section shape on sediment transport in open channel flow. Safari *et al.* [35] performed experiments in trapezoidal cross-section channel and taking data of various cross-sections from the literature developed general self-cleaning model applicable to drainage system design. As extension of previous studies, a channel cross-section shape factor is developed and incorporated in the model. Safari *et al.* [35] suggested:

$$\frac{V_n}{\sqrt{gd\,(s-1)}} = 7.34 C_v^{0.13} D_{gr}^{-0.12} \left(\frac{d}{R} \right)^{-0.44} \beta^{-0.91} \tag{6.5}$$

where β is channel cross-section shape factor defined as:

$$\beta = \frac{\sqrt{P/B}}{1.31 \, (B/D_h)^{-0.49}} \tag{6.6}$$

where P is wetted perimeter, B the water surface width and D_h the hydraulic depth of flow. The channel cross-section shape factor describes the effect of nonuniform distribution of shear stress on the channel boundary. More details about the self-cleaning studies in the literature are given in Safari *et al.* [18].

The mentioned studies above developed the models applying multiple nonlinear regression techniques. The left-hand side of self-cleaning models is particle Froude number (Fr_p) which gives the variables of flow mean velocity and sediment median size. It is seen that models developed by means of four fundamental characteristics of flow, fluid, sediment and channel.

6.1.2 Overview of machine learning application in drainage systems design

Machine learning techniques are known as powerful method for modeling of a variety of engineering problems. Hydraulics of sediment transport has complicated nature and developing analytical model to use in practice is a quite difficult task. To this end, as mentioned before, multiple nonlinear regression technique was mostly used for modeling sediment transport in drainage systems. Recently, applicability of different machine learning techniques were studied for designing of self-cleaning drainage system design. As primary studies, Ab Ghani and Azamathulla [36] and Azamathulla *et al.* [37] developed gene expression programming and adaptive neuro-fuzzy inference systems models considering the dimensionless parameters used in conventional regression models. Utilizing mostly the experimental data of Ab Ghani [29] and Vongvisessomjai *et al.* [5] experimental data, but not limited to them, numerous machine learning models were developed. As examples from the literature, particle swarm optimization, radial basis function neural network, wavelet-support vector machines (wavelet-SVMs), neuro-fuzzy-based group method of data handling, DT and artificial neural network (DT–ANN), combination of SVM with firefly algorithm, evolutionary polynomial regression, extreme learning machine, evolutionary algorithm with ANN, multigene genetic programming (MGGP), ANN, DT and multivariate adaptive regression splines are used for modeling of sediment transport in drainage systems in the literature ([38–42]; [43–51]).

As a novel technique for modeling sediment transport in drainage systems, this study provides EDT model through combination of DT and GP to enhance the model performance.

6.2 Materials and methods

6.2.1 Experimental data

Drainage systems and sewer pipes are classified as rigid boundary channels. Experimental data collected from rigid boundary open channels are used for modeling in this study. Mayerle [3], May [25], Ab Ghani [29] and Vongvisessomjai *et al.* [5] data sets are used in this study for modeling of sediment transport in drainage systems. Mayerle [3] performed experiments in circular and rectangular cross-section channels. Rectangular channels had width of 311 and 462 mm, and circular channel had diameter of 152 mm. Sediment particles in the range of 0.5–8.74 mm were used in the experiments. May [25] conducted experiments in a circular pipe with diameter of 450 mm using sediment with size of 0.73 mm. Ab Ghani [29] studied sediment transport in three different size pipes with diameters of 154, 305 and 405 mm using sediment size ranging from 0.46 to 8.3 mm. Vongvisessomjai *et al.* [5] performed experiments in two pipes having diameter of 100 and 150 mm using sediment size in range of 0.2–0.43 mm. Experimental data ranges are given in Table 6.3 and more details can be found in Safari *et al.* [18].

Table 6.3   Ranges of experimental data

Reference	D or W (mm)	d (mm)	λ	Y (mm)	V (m/s)	C_v (ppm)
Mayerle [3]	D = 152	0.50–8.74	0.016–0.034	28–122	0.37–1.10	20–1,275
	W = 311–462	0.50–5.22	0.011–0.025	31–111	0.41–1.04	14–1,568
May [25]	D = 450	0.73	0.014–0.018	222–338	0.50–1.22	2–38
Ab Ghani [29]	D = 154–450	0.46–8.30	0.013–0.048	24–342	0.24–1.21	4–1,450
Vongvise-ssomjai et al. [5]	D = 100–150	0.20–0.43	0.034–0.053	30–60	0.24–0.63	4–90

(D: circular channel diameter; W: rectangular channel bed width; d: sediment median size; λ: channel friction factor; Y: flow depth; V: flow mean velocity; C_v: sediment volumetric concentration).

6.2.2   Data preparation

With respect to the literature of sediment transport in rigid boundary channels, it is understood that a reliable model should incorporate four fundamental flow, fluid, sediment and channel characteristics. The basic variables of sediment transport in rigid boundary channels can be considered as flow mean velocity (V), hydraulic radius (R), gravitational acceleration (g), fluid specific mass (ρ) and kinematic viscosity (v), sediment median size (d), sediment volumetric concentration (C_v), sediment-specific mass (ρ_s) and channel friction factor (λ). Among these, V, R and g are flow characteristics, ρ and v as fluid, d, C_v and ρ_s as sediment, and λ as channel characteristic. Regarding the sediment transport models in rigid boundary channels, the aforementioned variables can be written as group of dimensionless parameters as follows:

$$\frac{V}{\sqrt{gd(s-1)}} = f\left(C_v, D_{gr}, \frac{d}{R}, \lambda\right) \tag{6.7}$$

Experimental data are re-formed based on the above equation to prepare the inputs and output of the model. It is seen that left-hand side of (6.7) as output of the model is particle Froude number, and parameters given at the right-hand side are inputs of the model. The prepared data are split in two parts for training and testing data sets. Through the training stage, the unknown relationship of input and output parameters is determined and in the testing stage the performance of the developed model is investigated on unseen data set. To this extent, among 375 data, 300 data for training and 75 data are randomly selected to test the models. Before the modeling, data are re-scaled to values between 0 and 1 to behave equally in the modeling procedure.

6.2.3   Performance indices

Performance evaluation of a developed model is of importance to understand its credibility. Hence, two statistical performance indices of root mean square error (*RMSE*) and concordance coefficient (*CC*) are used in this study. The *RMSE* measures the difference between the measured and calculated particle Froude numbers and is expressed as:

$$RMSE = \sqrt{\frac{\sum_{i=1}^{n} \left(Fr_p^m - Fr_p^c\right)^2}{n}} \qquad (6.8)$$

where Fr_p^m and Fr_p^c are the measured and calculated particle Froude numbers, respectively, while n denotes the number of observations. CC computes the concordance between the measured data and model output having ranges from 1 to -1 with the best value of 1. CC is computed as:

$$CC = \frac{2r\sigma_m\sigma_c}{\sigma_m^2 + \sigma_c^2 + \left(\overline{Fr_p^m} - \overline{Fr_p^c}\right)^2} \qquad (6.9)$$

where σ_m is standard deviation of the measured particle Froude numbers, σ_c denotes the standard deviation of the calculated particle Froude numbers, r stands for the correlation coefficient, $\overline{Fr_p^m}$ is the average of measured particle Froude numbers and $\overline{Fr_p^c}$ corresponds to the average of the calculated particle Froude numbers.

6.2.4 *Decision tree*

As a decision support system machine, DT is applied commonly in many engineering fields for classification and prediction problems [51,52]. DT utilizes the information available in the data set for classification purpose. With respect to the initial information on each class data, a tree-based model is generated as a diagram called DT. DT is a nonparametric regression method in which training algorithm output is the type of discrete classes. In order to obtain accurate results, entropy theory is used for characterization of impurity on a data set. DT generates discrete response for the model through an entropy reduction at its structure. In modeling procedure, a binary tree is used for finding the optimum tree size. The degree of importance of variables in DT is related to their location at DT flowchart, where preceding sets have higher entropy. Having the lowest entropy in leaves, makes it a confident decision which is based on the decision support system machine [53].

DT finds relationship between dependent and independent variables in nonlinear manner by series of binary splits. Assuming a sample X containing a set of independent variables $(x_1, x_2, \ldots, x_n)$, and dependent variable of y, DT determines y via subdivision of sample X. The tree is grown through successive branches of subsets of sample X into subsequent branches. The root node is defined as X and the binary tree is finalized in the set of terminal nodes. Based on independent variable vector value, the branches are determined at each node of binary tree. The branches selection at each parent node is based on the DT algorithm in which model variance in y has the maximum reduction. The decision at each node comes from considering a specific value of independent variable at the data range domain. Decisions stop once the model variance reduction reaches a specific value called critical number. Finally, in order to find the optimal tree size, the cross validation is used (Breiman *et al.*, 1987; [67]).

Figure 6.1 shows the DT flowchart for particle Froude number estimation. The effectiveness of the variables is related to the top-to-down induction of DT. It is seen from Figure 6.1 that dimensionless grain size parameter is selected as the most important parameter in particle Froude number estimation in DT technique. It can be linked to the fact that larger sediment particles need higher hydrodynamic forces to keep particle in motion within the flow. At second level of the DT flowchart, relative particle size is appeared, which indicates again the importance of the particle size behind hydraulic radius as flow characteristic. Sediment volumetric concentration is found as an effective parameter in next level that indicates particle Froude number is also more affected by sediment volumetric concentration. The interpretation given herein shows that sediment characteristics are most effective parameters in the Froude number estimation in DT technique.

6.2.5　GP model

GP is the generalization of 40-year-old genetic algorithm. Unlike genetic algorithm where the well-known Darwinian evolutionary process is used to optimize a function with predefined structure, GP is a symbolic regression technique that uses the same process to simultaneously develop both the optimal structure of the solution (linear or nonlinear) and its coefficients. In other words, the functional form of the solution has not chosen a priory. In the classic GP, the evolutionary process includes reproduction, crossover and mutation operations that act on the population of randomly created GP trees [55]. An example of randomly generated GP tree and its functional expression were illustrated in Figure 6.2. As shown in the figure, the individual has three levels (tree depth $= 3 + 1 = 4$) constructed from a root-node level (plus function at top), three inner nodes and five terminal nodes comprising two variables x_1 and x_2 and two random constants linked via branches.

During the evolutionary process, the best individual is transferred to the next population without any modification that is called reproduction. Crossover is the exchange of tree materials (nodes/branches) between desired individuals (called parents), and mutation is the substituting a randomly selected node (inner or terminal) with another node from the initial population. The process is iterated until the best suitable individual for the given problem is generated in which fitness function satisfies its criteria. From the predictive modeling perspectives, the main advantages of GP are summarized as follows:

- Like DT, GP provides the predictive model in a form of a tree or a mathematical equation that can be interpreted by modelers. The mathematical representation gives insights to the problem at hand and provides a great benefit in empirical modeling of unknown phenomena.
- GP is structure free. Unlike the traditional regression methods in which a pre-assumed functional form of the model is specified, GP evolves both functional form of the model and its optimum coefficients/parameters.

For details about GP, its different variants and applications in hydro-environmental studies, the interested reader is referred to a comprehensive review paper published by Danandeh Mehr *et al.* [56].

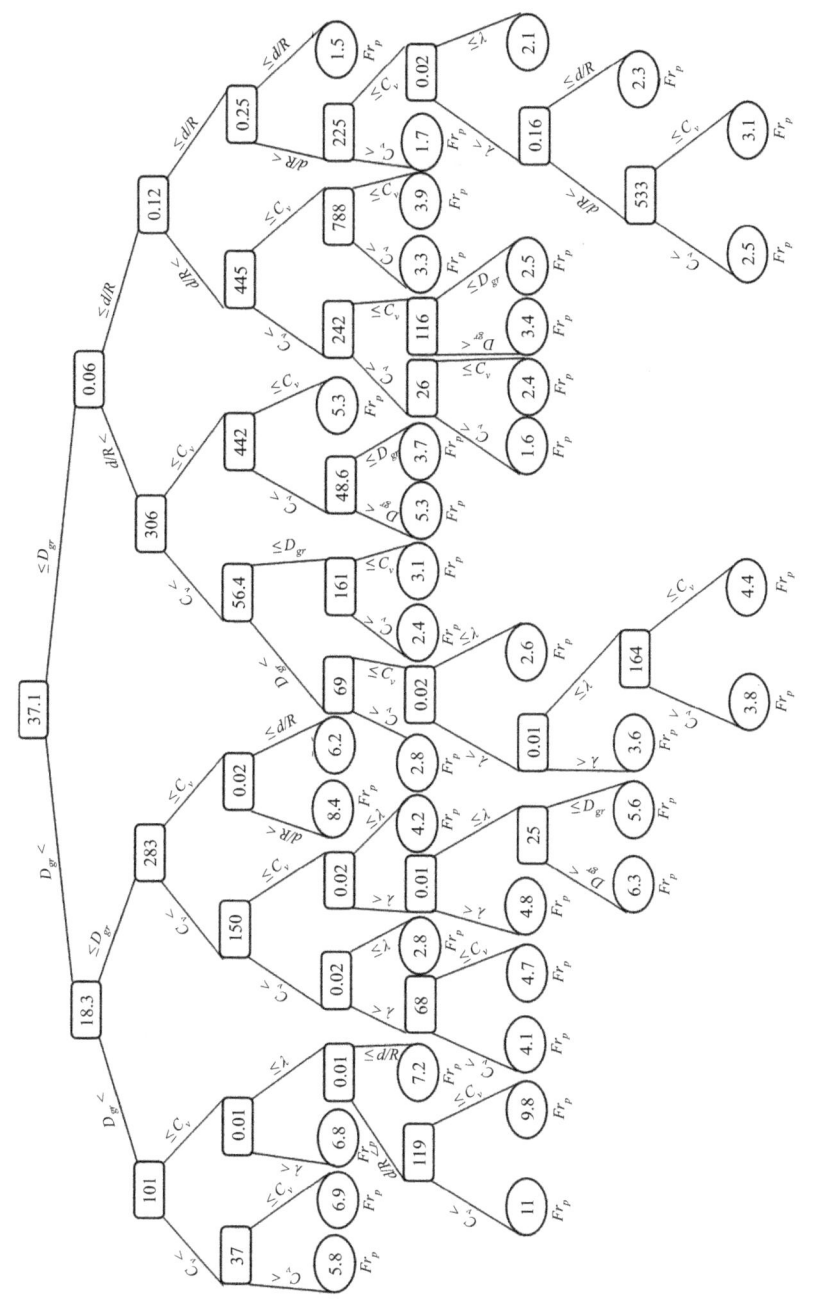

Figure 6.1   DT flowchart for particle Froude number estimation

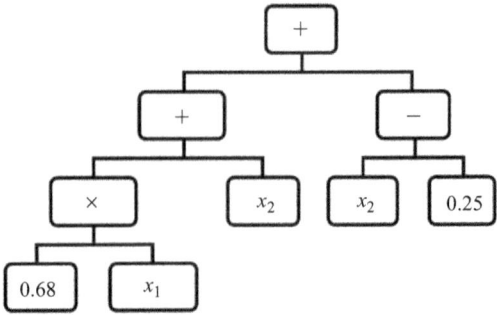

Figure 6.2   Structure of GP tree representing the function $y = 0.68x_1 + x_2 + (x_2 - 0.25)$

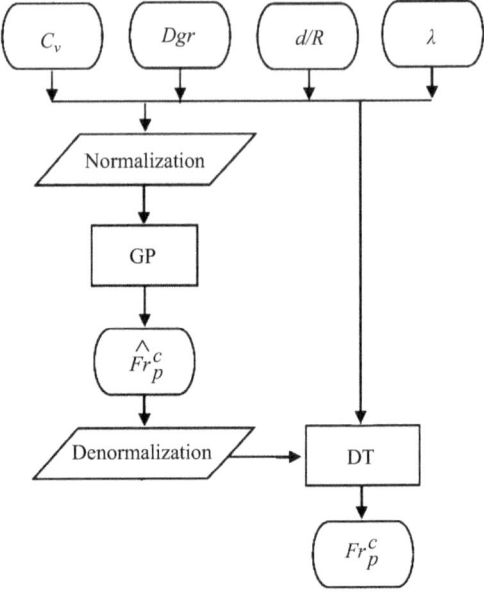

Figure 6.3   EDT model for particle Froude number prediction

6.2.6   Hybrid EDT model

Considering the strengths of GP and limitations of DT, integrating GP and DT was considered in this study for the development of final sediment deposition velocity formulae. By this way, the calculated particle Froude numbers (Fr_p^c) obtained by GP was considered as the auxiliary input for DT-based modeling. Figure 6.3 illustrates the proposed EDT model for Fr_p^c prediction.

The proposed EDT model is a hybrid evolutionary model that intends for fine-tuning of the DT inputs attained from the canonical GP. While stand-alone GP/DT typically tries to improve the predictive accuracy of the best solutions through increasing their tree elements, the proposed EDT model integrates a low complex GP solution ($\widehat{Fr_p^c}$) with nonlinear DT technique. As illustrated in Figure 6.3, the proposed model includes three phases of data preparation, GP modeling and DT mapping. In the first phase, the input parameters ($C_v, D_{gr}, \frac{d}{R}, \lambda$) are normalized using well-known min-max normalization procedure. The result is dimensionless inputs within the range [0–1] that is required for dimensionally aware GP solution. Then, denormalized GP solution in alliance with the same GP inputs is entered to DT to generate a population of suitable predictive models with limited maximum allowable tree depth to avoid from the generation of complex models.

6.2.7   Comparison of models

The DT, GP and EDT models developed in this study are compared with selected three regression models of Mayerle *et al.* [28], Ab Ghani [29] and Vongvisessomjai *et al.* [5] in terms of *RMSE* and *CC* in Table 6.4 on testing data set. It is seen from Table 6.4 that three machine learning techniques of DT, GP and EDT outperform regression models. Although the machine learning techniques provide similar results with a small differences in terms of *RMSE* and *CC*, EDT method, which is the combination of DT and GP techniques, outperforms the stand-alone methods. EDT with *RMSE* and *CC* of 0.93 and 0.91 is found superior to its alternatives in particle Froude number prediction. GP gives better results than DT in terms of *RMSE* and *CC*. Vongvisessomjai *et al.*'s [5] results are better than those of Ab Ghani [29] and Mayerle *et al.* [28] models; however, its accuracy is not as high as in DT, GP and EDT models.

Comparison of measured and calculated particle Froude numbers on testing data set for three machine learning techniques of DT, GP and EDT with those of Mayerle *et al.* [28], Ab Ghani [29] and Vongvisessomjai *et al.* [5] regression models is shown in Figure 6.4. It is illustrated in Figure 6.4 that DT provides best results for particle Froude numbers less than 7 where the data fall on the best fit line; however, for particle Froude numbers higher that 7, a large scatter is shown in DT performance. GP provides acceptable results where the data are close to the best fit line. EDT as combination of the DT and GP techniques improves the performance of DT significantly for particle Froude numbers higher than 7. Regression models of Mayerle *et al.* [28], Ab Ghani [29] and Vongvisessomjai *et al.* [5] considerably overestimate particle Froude number, although Ab Ghani [29] and Vongvisessomjai *et al.* [5] outperform Mayerle *et al.* [28] model.

Conventional drainage system design method based on unique value of velocity or shear stress is not recommended due to the neglecting the effective variables in hydraulics of the sediment transport as shown in Tables 6.1 and 6.2. To this extent, regression models were developed to incorporate more hydraulic parameters. However, regression techniques have no enough capability to calculate precisely. It is the main motivation of the applying machine learning techniques for modeling of sediment transport in open channel flow. Additional to the applying of the three robust

Table 6.4   Performance of the models in terms of RMSE and CC

Method	RMSE	CC
Machine learning models		
DT	1.03	0.86
GP	0.96	0.89
EDT	0.93	0.91
Conventional regression models		
Mayerle *et al.* [28]	3.14	0.59
Ab Ghani [29]	1.51	0.79
Vongvisessomjai *et al.* [5]	1.27	0.84

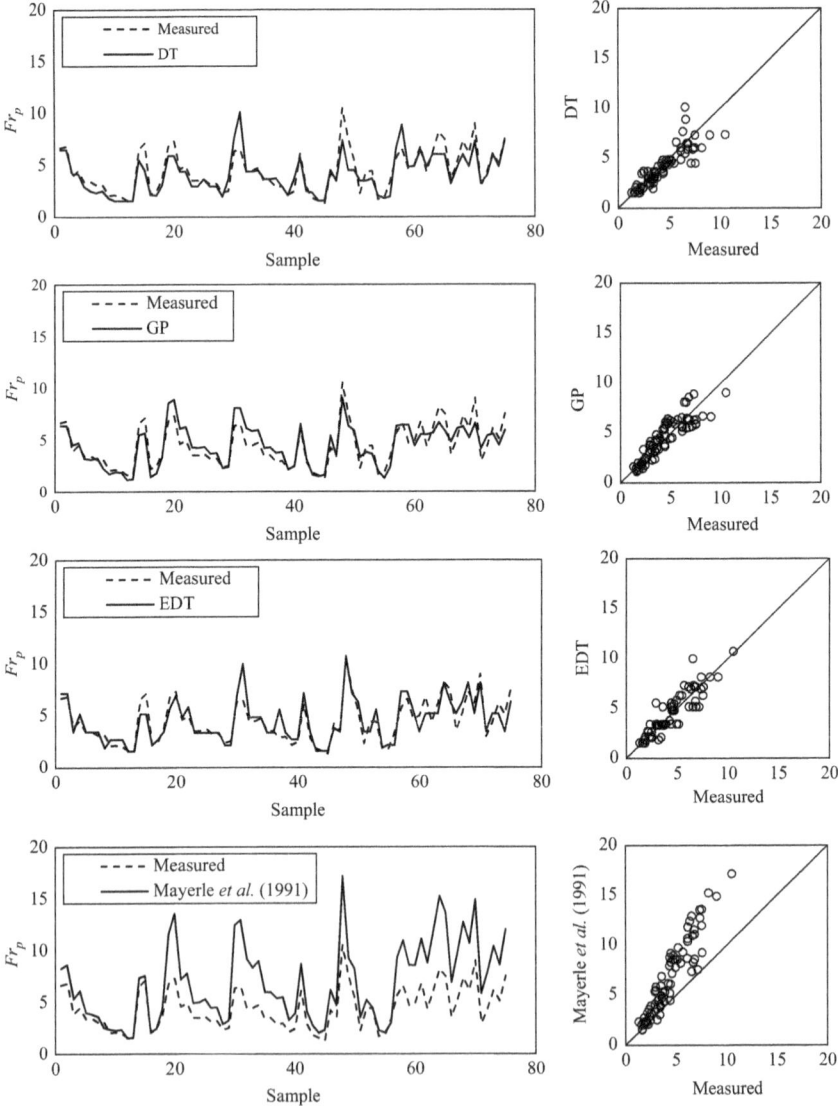

Figure 6.4   Comparison of measured and calculated Fr$_p$

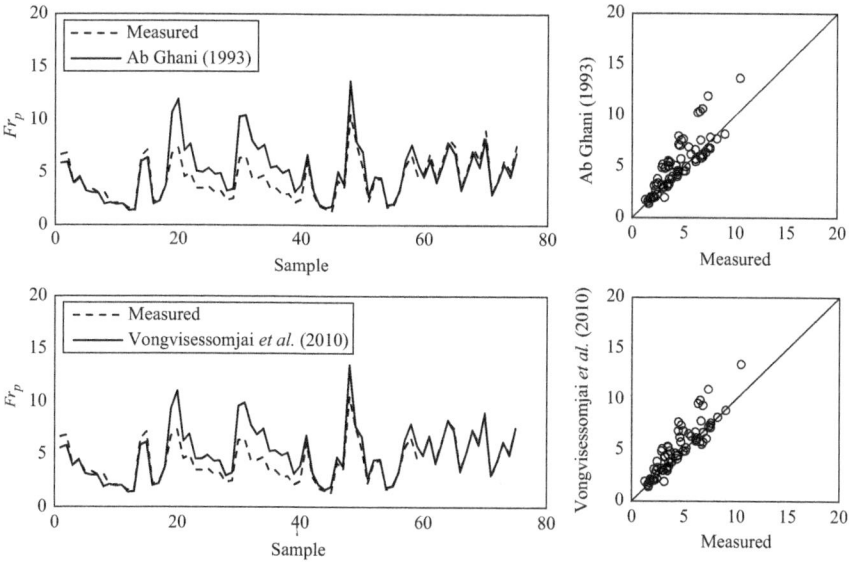

Figure 6.4   Continued

techniques of DT, GP and EDT for modeling of the sediment transport in drainage systems, this study used wide range of experimental data which can be considered as reliable models to use in practice.

6.2.8   *IoT applications in drainage systems*

A brief review of the applications of IoT in hydro-environmental engineering was recently provided by Nourani *et al.* [57] and Al-Turjman [58]. The authors also concluded that only a few studies investigated use of IoT in hydro-environmental issues relevant to the smart city concept. Currently, smart city is growing through environmental monitoring applying the IoT. In order to make smart city, smart water meters and smart drainage systems must be considered [59]. It is an essential issue to design drainage channels to work appropriately to keep urban area clean and healthy [60]. IoT is defined as a concept to enhance reliability by self-managing operational performance [61,62]. The smart cities are equipped using a variety of electronic apparatus such as sensors and actuators which are used widely in IoT technology [62]. The IoT has an advantage to restrain the real-time adverse environmental problems, which simplifies the self-managing operational of the system ([63]). As an efficient tool for proactively capturing of any destruction in the system, the mounted devices send alert by real-time monitoring of the system [61]. IoT structure includes three layers of sensing, network and application. Electronic devices such as sensors and actuators are positioned in the sensing layer. Through utilizing one of the Internet technologies of Wi-Fi, 2G, 3G, 4G and power-line communication as wireless transmitters, data are collected and sent to the application layer. Within the application layer, the data are received and processed [62,64].

The IoT technology is used in smart cities to monitor the urban drainage channel performance to generate data for system maintenance [64,65]. It is quite difficult task to figure out the blockage location in conventional drainage channels, and the maintenance of blocked channel is a time-consuming task. The application of the IoT in monitoring of drainage systems is considered already for determination of the exact blockage location along the pipes and more importantly for receiving early alarm before blockage in smart city. Drainage channel blockage has several environmental and economical adverse effects. To this end, sensors are deployed along the pipes to detect the flow condition to protect and repair the pipes before complete blockage [66].

Prediction of sediment deposition and channel blockage in advance seem to be reasonable instead of detecting blockage location using IoT technology. It helps for a better decision-making and efficiencies. Therefore, for optimizing the applicability of the IoT technology, it can be said that detecting the flow condition in which sediment start to deposit would be helpful to satisfy a nondeposition condition. Designing the drainage channels based on hydraulic point of view and considering fundamental flow, fluid, sediment and channel characteristics to understand the flow velocity at sediment deposition condition seems to have more scientific justification. To this extent, positioned sensors along the drainage pipe must detect and read the velocity continually and alert once the flow velocity decreases to a critical value of sediment deposition. This scenario is suggested in this study for monitoring of urban drainage channels in smart cities. It has advantages such as making city clear and intelligent management of sewer and drainage systems through unendingly monitoring of flow condition and automatic sending of the flow rate and flow velocity via wireless transmitters system to the application layer.

6.3   Conclusions

In order to have clean, safe and healthy smart city, it is an essential issue to consider smart drainage systems. The first application of the IoT technique in drainage systems is used to detect the location of the sediment deposition and blockage. From the hydraulic point of view, this study recommends a novel scenario for monitoring the drainage channels in which the mounted sensors must detect and record the flow rate and flow velocity and alert once flow reaches the critical sediment deposition velocity. For determination of deposition condition of the sediment in drainage systems, this study applied three robust machine learning techniques of the DT, GP and EDT for modeling of sediment transport within the flow. The results show the superiority of the EDT model which is developed through the combination of the DT and GP stand-alone techniques. Comparison of the models in terms of variety of statistical performance indices shows that DT, GP and EDT machine learning techniques outperform regression models. The best performance of the developed models in this study can be attributed to the applying techniques as well as utilizing wide ranges of the experimental data. Results of this study can be considered as economical and flexible solution for urban derange system management in smart city.

References

[1]   CIRIA. 1986. Sediment movement in combined sewerage and storm-water drainage systems. Phase 1. Project Report. London: CIRIA Research Project No. 336.

[2]   Safari, M. J. S. 2016. Self-cleansing drainage system design by incipient motion and incipient deposition-based models. PhD Thesis, Istanbul Technical University, Turkey.

[3]   Mayerle, R. 1988. Sediment transport in rigid boundary channels. PhD Thesis, University of Newcastle upon Tyne, UK.

[4]   Nalluri, C., and Ab Ghani, A. 1996. Design options for self-cleansing storm sewers. Water Sci. Technol. 33(9), 215–220.

[5]   Vongvisessomjai, N., Tingsanchali, T., and Babel, M. S. 2010. Non-deposition design criteria for sewers with part-full flow. Urban Water J. 7(1), 61–77.

[6]   ASCE Water Pollution Control Federation. 1970. Design and construction of sanitary and storm sewers. Rep. No. 37, ASCE, N.Y.

[7]   British Standard Institution. 1987. Sewerage. Guide to new sewerage construction. BS8005 Part 1.

[8]   Minister of Interior. 1977. Instruction technique relative aux re'seaux d'assainissement des agglomerations. Minister of interior, circulaire interministerielle. IT 77284 INT. Paris, France.

[9]   European Standard EN 752-4. 1997. Drain and sewer system outside building: Part 4. Hydraulic design and environmental considerations. Brussels: CEN (European Committee for Standardization).

[10] Arbeitsblatt ATV-A 110, 1998. Richtlinien für die hydraulische Dimensionierung und den Leistungsnachweis von Abwasserkanälen und -leitungen. Hennef, GFA e.V.

[11] ATV-DVWK-Regelwerk, 2001. Hydraulische Dimensionierung und Leistungsnachweis von Abwasserkanälen und -leitungen. Hennef, GFA e.V.

[12] DID. 2012. Urban stormwater management manual for Malaysia. 2nd edition, Kuala Lumpur, Malaysia: Department of Irrigation and Drainage, Malaysia.

[13] Lysne, D. K. 1969. Hydraulic design of self-cleaning sewage tunnels. Journal of the Sanitary Engineering Division of the American Society of Civil Engineers 95(SA1), 17–36.

[14] Yao, K. M. 1974. Sewer line design based on critical shear stress. Journal of the Environmental Engineering Division of the American Society of Civil Engineers 100(EE2), 507–520.

[15] Lindholm, O. G. 1984. Pollutant loads from combined sewer systems. In: Proceedings of the 3rd International Conference on Urban Storm Drainage, Gothenburg, Sweden, June, Vol. 4, 1602.

[16]   Scandiaconsult. 1974. Synopsis of research programme for self-cleansing sewers. Stockholm: Orrje.

[17]   Brombach, H., Michelbach, S., and Wohrle, C. 1992. Sedimentation- und Remobilisierungsvorgange NEIDERSCHLAG, Eigenverlag Umwelt- und fluid- technik GmbH, Bad Mergentheim.

[18]   Safari, M. J. S., Mohammadi, M., and Ab Ghani, A. 2018. Experimental studies of self-cleansing drainage system design: A review. J. Pipeline Syst. Eng. 9(4), 04018017.

[19]   Novak, P., and Nalluri, C. 1975. Sediment transport in smooth fixed bed channels. J. Hydraul. Div. ASCE 101 (HY9), 1139–1154.

[20]   Novak, P., and Nalluri, C., 1984. Incipient motion of sediment particles over fixed beds. J. Hydraul. Res. 22 (3), 181–197.

[21]   Macke, E. 1982. About sedimentation at low concentrations in partly filled pipes. PhD Thesis, Technical University of Braunschweig, Germany.

[22]   Arora, A.K. 1983. Velocity distribution and sediment transport in rigid bed open channels. PhD Thesis, University of Rookee, Rookee, India.

[23]   Nalluri, C., and Spaliviero, F. 1998. Suspended sediment transport in rigid boundary channels at limit deposition. Water Sci. Technol. 37 (1), 147–154.

[24]   May, R. W. P., Brown, P. M., Hare, G. R., and Jones, K. D. 1989. Self-cleansing conditions for sewers carrying sediment. Report SR, 221.

[25]   May, R. W. P. 1993. Sediment transport in pipes and sewers with deposited beds. Technical Report, Hydraulic Research Ltd., Report SR 320, Wallingford, UK.

[26]   Ackers, P., and White, W.R. 1973. Sediment transport: New approach and analysis. J. Hydr. Eng. Div. ASCE, 99 (HY11), 2041–2060.

[27]   Ackers, P. 1984. Sediment transport in sewers and the design implications. Intern. Conf. on Planning, Construction, Maintenance, and Operation of Sewerage Systems, BHRA/WRc, Reading, England, pp. 215–230.

[28]   Mayerle, R., Nalluri, C., and Novak, P. 1991. Sediment transport in rigid bed conveyances. J. Hydraul. Res. 29 (4), 475–495.

[29]   Ab Ghani, A. 1993. Sediment transport in sewers. PhD Thesis, University of Newcastle upon Tyne.

[30]   Butler, D., May, R., and Ackers, J. 2003. Self-cleansing sewer design based on sediment transport principles. J. Hydraul. Eng. 129 (4), 276-282.

[31]   De Sutter, R., Rushforth, P., Tait, S., Huygens, M., Verhoeven, R., and Saul, A. 2003. Validation of existing bed load transport formulas using in-sewer sediment. J. Hydraul. Eng. 129 (4), 325–333.

[32]   May, R. W., Ackers, J. C., Butler, D., and John, S. 1996. Development of design methodology for self-cleansing sewers. Water Sci. Technol. 33 (9), 195–205.

[33]   Ota, J. J. 1999. Effect of particle size and gradation on sediment transport in storm sewers. PhD Thesis, University of Newcastle Upon Tyne, UK.

[34]   Ota, J. J., and Perrusquia, G. S. 2013. Particle velocity and sediment transport at the limit of deposition in sewers. Water Sci. Technol. 67(5), 959–967.

[35]   Safari, M. J. S., Aksoy, H., Unal, N. E., and Mohammadi, M. 2017. Non-deposition self-cleansing design criteria for drainage systems. J. Hydro-environ. Res. 14, 76–84.

[36]   Ab Ghani, A., and Azamathulla, H. M. 2010. Gene-expression programming for sediment transport in sewer pipe systems. Journal of Pipeline Systems Engineering and Practice 2(3), 102–106.

[37]   Azamathulla, H. M. D., Ab Ghani, A., and Fei, S. Y. 2012. ANFIS-based approach for predicting sediment transport in clean sewer. Appl. Soft Comput. 12, 1227–1230.

[38]   Ebtehaj, I., Bonakdari, H., Shamshirband, S., and Mohammadi, K. 2015. A combined support vector machine-wavelet transform model for prediction of sediment transport in sewer. Flow Meas. Instrum. 47, 19–27.

[39]   Ebtehaj, I., Bonakdari, H., and Zaji, A. H., 2016. An expert system with radial basis function neural network based on decision trees for predicting sediment transport in sewers. Water Sci. Technol. 74 (1), 176–183.

[40]   Ebtehaj, I., Bonakdari, H., Shamshirband, S., Ismail, Z., and Hashim, R. 2016. New approach to estimate velocity at limit of deposition in storm sewers using vector machine coupled with firefly algorithm. J. Pipeline Syst. Eng. 8(2), 04016018.

[41]   Safari, M. J. S., Aksoy, H., and Mohammadi, M. 2016. Artificial neural network and regression models for flow velocity at sediment incipient deposition. J. Hydrol. 541, 1420–1429.

[42]   Safari, M. J. S., Shirzad, A., and Mohammadi, M. 2017. Sediment transport in deposited bed sewers: Unified form of May's equations using the particle swarm optimization algorithm. Water Sci. Tech. 76 (4), 992–1000.

[43]   Ebtehaj, I., and Bonakdari, H. 2016. Assessment of evolutionary algorithms in predicting non-deposition sediment transport. Urban Water J. 13, 499–510.

[44]   Najafzadeh, M., and Bonakdari, H. 2016. Application of a neuro-fuzzy GMDH model for predicting the velocity at limit of deposition in storm sewers. J. Pipeline Syst. Eng. 8, 06016003.

[45]   Najafzadeh, M., Laucelli, D. B., and Zahiri, A. 2017. Application of model tree and evolutionary polynomial regression for evaluation of sediment transport in pipes. KSCE J. Civ. Eng. 21, 1956–1963.

[46]   Roushangar, K., and Ghasempour, R. 2017. Prediction of non-cohesive sediment transport in circular channels in deposition and limit of deposition states using SVM. Water Sci. Tech.- W. Sup. 17, 537–551.

[47]   Roushangar, K., and Ghasempour, R. 2017. Estimation of bedload discharge in sewer pipes with different boundary conditions using an evolutionary algorithm. Int. J. Sediment Res. 32 (4), 564–574.

[48]   Safari, M. J. S., and Danandeh Mehr, A. 2018. Multigene genetic programming for sediment transport modeling in sewers at non-deposition with deposited bed condition. Int. J. Sediment Res. (33) 3, 262–270.

[49]   Wan Mohtar, W. H. M., Afan, H., El-Shafie, A., Bong, C. H. J., Ab. and Ghani, A. 2018. Influence of bed deposit in the prediction of incipient sediment motion in sewers using artificial neural networks. Urban Water J. 15 (4), 296–302.

[50]   Safari, M. J. S., and Shirzad, A. 2019. Self-cleansing design of sewers: Definition of the optimum deposited bed thickness. Water Environ. Res. 91(5), 407–416.

[51]   Safari, M. J. S. 2019. Decision tree (DT), generalized regression neural network (GR) and multivariate adaptive regression splines (MARS) models for sediment transport in sewer pipes. Water Sci. Technol. 79(6), 1113–1122.

[52]   Quinlan, J. R. 1986. Induction of decision trees. Machine Learning 1, 81–106.

[53]   Vaheddoost, B., Aksoy, H., Abghari, H., and Naghadeh, S. 2015. Decision tree for measuring the interaction of hyper-saline lake and coastal aquifer in Lake Urmia. In Proceeding of Environmental and Water Resource Institute (EWRI): Watershed Management Symposium, August, pp. 5–7.

[54]   Balk, B., and Elder, K. 2000. Combining binary decision tree and geostatistical methods to estimate snow distribution in a mountain watershed. Water Resour. Res. 36(1), 13–26.

[55]   Danandeh Mehr, A. 2018. An improved gene expression programming model for streamflow forecasting in intermittent streams. J. Hydrol. 563, 669–678.

[56]   Danandeh Mehr, A., Nourani, V., Kahya, E., Hrnjica, B., Sattar, A. M., and Yaseen, Z. M. 2018. Genetic programming in water resources engineering: A state-of-the-art review. J. Hydrol., 566, 643–667.

[57]   Nourani V., Molajou A., Najafi H., and Danandeh Mehr A. 2019 Emotional ANN (EANN): A new generation of neural networks for hydrological modeling in IoT. In: Al-Turjman F. (eds) Artificial Intelligence in IoT. Transactions on Computational Science and Computational Intelligence. Springer, Cham.

[58]   Al-Turjman, F. 2019. The road towards plant phenotyping via WSNs: An overview. Elsevier Comput. Electron. Agr. 161, 4–13.

[59]   Al-Turjman, F. 2019. 5G-enabled devices and smart-spaces in social-IoT: An overview. Elsevier Future Generation Computer Systems 92(1), 732–744.

[60]   Wang, S., Zhang, Z., Ye, Z., Wang, X., Lin, X., and Chen, S. 2013. Application of environmental Internet of Things on water quality management of urban scenic river. International Journal of Sustainable Development & World Ecology 20(3), 216–222.

[61]   Edmondson, V., Cerny, M., Lim, M., Gledson, B., Lockley, S., and Woodward, J. 2018. A smart sewer asset information model to enable an "Internet of Things" for operational wastewater management. Automation in Construction, 91, 193–205.

[62]   Talari, S., Shafie-Khah, M., Siano, P., Loia, V., Tommasetti, A., and Catalão, J. 2017. A review of smart cities based on the Internet of Things concept. Energies, 10(4), 421.

[63]   Al-Turjman, F., Altrjman, C., Din, S., and Paul, A. 2019. Energy monitoring in IoT-based ad hoc networks: An overview. Elsevier Computers & Electrical Engineering Journal 76, 133–142.

[64]   Zhang, D., Lindholm, G., and Ratnaweera, H. 2018. Use long short-term memory to enhance Internet of Things for combined sewer overflow monitoring. J. Hydrol. 556, 409–418.

[65]   Montserrat, A., Bosch, L., Kiser, M. A., Poch, M., and Corominas, L. 2015. Using data from monitoring combined sewer overflows to assess, improve,

and maintain combined sewer systems. Science of the Total Environment, 505, 1053–1061.

[66]   Drenoyanis, A., Raad, R., Wady, I., and Krogh, C. 2019. Implementation of an IoT based radar sensor network for wastewater management. Sensors, 19(2), 254.

[67]   Breiman, L., Friedman, J., Olshen, R., and Stone, C. 1987. Classification and Regression Trees. Wadsworth, Belmont, Calif.

Chapter 7

Statistical analysis for sensory E-Health applications: opportunities and challenges

Hakan Yekta Yatbaz[1] and Ali Cevat Taşıran[2]

Population of people whose age are over 60 is expected to be more than double by 2050. As they live alone or without any healthcare professional, it is a complicated job to understand emergency situations or mild symptoms of the diseases. On the other hand, smart home environments offer an unrivalled opportunity to collect data for monitoring to understand human behaviours so that healthcare institutions intervene such cases. Furthermore, thanks to the availability of large smart home data sets on the Internet with different features such as different number of residents, house plans and different use cases (activities of daily living (ADL), work activities) [1,2], it is possible to work on different aspects in this area. Hence, forthcoming smart cities can highly benefit from those electronic health (E-Health) systems to enhance Ambient Assisted Living (AAL). This chapter presents the opportunities and the challenges that come with this type of data and the applications while emphasizing the importance of statistical analysis on these topics. In addition, it shows the steps of analysis of the data from binary sensors deployed in a smart home.

List of abbreviations

ADL   Activities of daily living
AAL   Ambient Assisted Living
ABS   Anonymous Binary Sensor
BCD   Behaviour change detection
AR   Activity recognition
DCNN   Deep convolutional neural network
LSTM   Long short-term memory
CRF   Conditional random field
DBN   Dynamic Bayesian network
WHO   World Health Organization
WSN   Wireless sensor network

[1]Computer Engineering Department, Middle East Technical University, Northern Cyprus Campus, Kalkanlı, Güzelyurt, Turkey
[2]Economics Department, Middle East Technical University, Northern Cyprus Campus, Kalkanlı, Güzelyurt, Turkey

7.1   Introduction

Increasing numbers of devices equipped with variety of sensors have started to take more and more part in the concept of Internet of Things (IoT), and these devices are being used in various application areas [3]. The global IoT market is nearly doubled in size between 2014 and 2017, passing the one trillion US dollars mark by that year [4]. Furthermore, making everything smart is also become popular with the vast opportunities provided by IoT applications. Eventually, these trends led society to think about smart cities with different smart environments, such as buses, houses and hospitals, that people encounter work collaboratively enough to make the city management automatized, i.e. smart, and quality of life better.

E-Health is one of the most important advancements over the existing healthcare systems since it introduces machine learning, wireless sensor networks (WSNs) and communications technology to the area to make healthcare systems more efficient and available at any time. According to [5], 70% of the member states of World Health Organization (WHO) have a national E-Health policy or strategy including ethical frameworks, funding strategies and education. With the increasing popularity and demand on E-Health, it becomes an essential module for smart cities to make healthcare systems automatized. WSNs have great importance on the improvement of E-Health applications as they enable researchers to build systems that can collect information from various distributed sources [6]. As WSN-based E-Health systems are providing large amount of data for processing and information extraction, it also enables researchers to work on big data to enhance the current early warning systems to provide better health services [7]. Furthermore, WSN is a popular area that is contributed heavily [8,9]. As the scope of the E-Health concept covers a variety of subjects such as assisted living, patient monitoring and early warning systems, the applications that are required for smart cities can be categorized differently.

AAL is a subfield of E-Health which aims to enable people to stay active longer and connected without requiring physical assistance by using the information and communication technologies [10]. The concept itself is essential for any smart city or home environment as the majority of the applications and research conducted in AAL focuses on elderly and disabled people to make their life more independent. Obviously, AAL systems need to interact actively or passively with their owner to collect data for further processing and analysing. These assisted living systems can be categorized based on the interaction type determined by the sensors used and their mobility. These categories are the following:

1. **Camera-based systems** [11,12]: Systems that recognize activities from a video or an image.
2. **Anonymous Binary Sensor (ABS)-based** systems [1,2,13]: Systems that recognize activities from binary sensors (motion) that collect data unobtrusively.
3. **Wearable device-based systems** [14,15]: Systems that recognize activities from wearable devices (gyroscope, accelerometer, ECG, etc.).

Considering the categories stated, a trade-off between accuracy and privacy can be reported since camera and wearable device-based systems can obtain personal and sensitive data frequently. However, since the data collected are sensitive, it creates problems in terms of privacy even though such systems have higher accuracy due to the extensive information they provide to the recognition system. On the other hand, ABS-based systems do not have the privacy issue since sensors used in these systems are collecting data; only there is a triggering action around them. However, since they are generating information about their on/off status, implementing recognition with such data does not give the same accuracy with the other categories. Moreover, another metric to compare these systems is their energy efficiency. Since the camera and wearable sensor-based applications include one or many wireless multimedia sensors such as camera or audio sensors, the data generated by those devices are considerably large compared to the binary sensors, so transmission of those data to the sink node requires more energy than binary sensors [16]. Therefore, the most preferred solution for such systems is to use binary sensors as they have no problem in terms of privacy with the elder people and it is a greener solution compared to the other methods. Nevertheless, there are not many models or algorithms to achieve high accuracy on data sets with binary sensor data. As the possible sensors and environments for AAL systems are vast, there are many opportunities and challenges for designing such systems as well as analysing the data gathered.

In this chapter, we focus on statistical analysis of an example data set of a smart home data deployed with motion sensors to understand the ADL of an elderly women. By explaining the need for both descriptive and inferential statistical methods, we show the information that can be extracted from such limited information to enhance the existing recognition algorithms.

The structure of this chapter is as follows. Section 7.2 examines the studies done on AAL with the data set selected in this study. Section 7.3 reports the challenges faced during pre-processing done on the data set, and opportunities and future directions are given in Section 7.4. The final section concludes this chapter.

7.2   Background

In the last decade, many data sets are presented for activity recognition (AR) using smart home system. Washington State University Center for Advanced Studies in Adaptive Systems (WSU CASAS) have around 70 open data sets with different number of residents, house plans and purposes to evaluate the performance of the proposed methods [17]. Many of the existing studies use Aruba, Tulum or Twor data sets from WSU CASAS [1,2,17], Kasteren data set [18], HMBS data set [19] or their own experiment specific data sets. Majority of the studies focuses on two main tasks.

Considering the studies used binary sensors, the trend topics can be summarized in two different categories:

1. behaviour change detection (BCD) and
2. activity recognition (AR).

In the first category, researchers examine the changes in the patterns of an activity after a health event, i.e. a disease or an accident occurrence. To do so, they extract the features such as the change in the total time spent or change in the number of instances of activities to understand behaviour change. In Sprint *et al.* [20], two different classifiers for BCD are compared on three different use cases. These changes are given as diagnosis with lung cancer, diagnosis with insomnia, falling in home. The change of the daily life activities they performed has been examined and it is showed that there is a significant change for each participant's activities. In addition, in [21], researchers tried to enhance the identification of health state changes by presenting a new method where an expert healthcare professional, an expert nurse, is added to the team to analyse the data along with the automated system and to give feedback about the changes. In this study, they emphasized the importance of motion patterns with their use case, the detection of restless leg syndrome from the data set. Furthermore, it is also shown that, an expert-in-the-loop system enables the automated recognition system to detect health state changes as a healthcare professional would. In Rantz *et al.* [22], the variety of sensors are increased. In their experiment, in addition to binary sensors they added some passive scalar sensors to detect the sleep quality, heart rate (from bed), etc., and they deployed their system to a nursing home. Since all the sensors they used are passive, they did not face any privacy problems with the residents. Using the data gathered from the sensors, and the collaborative analysis of the data with the system and healthcare professionals similar to the study in [21], they managed to recognize the urinary tract infections despite some false positive alerts.

In the second category, the main focus is not to detect the changes on the activities but to be able to find the activities in an unannotated data set with the help of the prior knowledge obtained from annotated data set. In their study in Gochoo *et al.* [13], to understand the activities of the elderly people living alone (elderly woman), they used a deep convolutional neural network (DCNN) for activity learning. They used the open Aruba data set [1] and proposed a new method for representing ABS logs called activity image which is a binary 2D image. To obtain this image, they used the y-axis as the indicator for the each sensor and x-axis as the indicator for ordered activities. In other words, in this binary image, white pixel with (x, y) coordinates indicates that the sensor with the ID at yth point is ON at xth point, and if the pixel is black, it means its OFF. In addition, they stated that to best of their knowledge their classifier yielded the best recognition rate among the other classifiers used on the data set. On the other hand, they indicated that improvement can be achieved on their classifier using long short-term memory (LSTM) model with DCNN to obtain real-life long-time monitoring system. Krishnan and Cook [23] and Cook *et al.* [24] have developed activity recognition models for the single resident environment using three data sets on different test beds with 32 binary sensors. All three data sets include 11 daily activities; 7 of them are the same as the data set mentioned in this chapter. In addition to the studies stated, the applications that are used statistical models to predict activity recognition are presented in Table 7.1.

In [26], a novel sliding window approach is used with two different classification methods for activity recognition. They introduced two different activity types named instantaneous and durational activities, and proposed separate classification techniques for each class. They used logistic regression for the durational activities

Table 7.1 Studies using statistical models for AR

References	Windowing method	Feature extraction	Classification method	Data set	Result
[25]	Activity based	Spatiotemporal	Coarse-grained and fine-grained	Aruba [1,2]	0.78
[26]	Time based	Temporal	Logistic regression	Aruba [1,2]	0.85
[27]	Time based	Spatial and temporal	DBN	Aruba [1,2]	0.416
[28]	Probabilistic dynamic sensor based	Spatial and temporal	Multinominal logistic regression	Aruba [1,2]	0.7014
[29]	Sensor based	Temporal and semantic	CRF	Aruba [1,2]	0.80
				Twor [1,2]	0.79
				Tulum [1,2]	0.75

which have more sensor events in each instance. Similarly, in [28], multinominal logistic regression is used along with a proposed probabilistic dynamic windowing for human activity recognition. In their study, they used Aruba data set [1,2], but they reported that they eliminated the Resperate and Wash_Dishes activities due to their low number of instances. On the other hand, they managed to include Housekeeping activity even though it has less number of instances than Wash_Dishes activity. Although they showed that their proposed windowing method is better than the existing methods, they also stated that the F measure of recognition without windowing is higher. In another study, a two-layer framework is presented for activity recognition. One main advantage of the proposed framework is its suitability for recognizing highly overlapping activities in small areas [25]. Their method first split the activity instances based on the feature vectors and activity phenomena matrix they proposed by checking whether they are highly overlapping or not. After separation, two different classification models are used for each category. Agarwal and Flach used conditional random fields (CRFs) to classify activity instances [29]. However, they stated that CRF is computationally expensive method in training part. On the other hand, they also stated that it is a fast method to classify after training is complete. Dynamic Bayesian network (DBN) is also used to classify activities and tested with Aruba data set to show its suitability in [27].

All in all, there are many different techniques and data sources to perform activity recognition for different purposes. However, one of the most popular data sources to do activity recognition is to use smart home systems with binary sensor data due to its cost efficiency and its unobtrusiveness. Studies examined in this section have variety of different windowing, feature extraction and classification methods with their results tested on different data sets. As presented in Table 7.1, majority of the studies used F_1 score as evaluation metric. On the other hand, some studies present metrics such as precision and recall where F_1 score can be calculated using these. The table presented shows the maximum F score values for the studies who proposed multiple methods, and displays mean F score values of activity classes in the data set used for the studies which have no F score value but present the precision and recall values for their experimentation whose formulas are given in the equations 7.1, 7.2 and 7.3. Furthermore, one of the most used data sets is Aruba data set where the lowest F score value obtained is 0.416 with DBN-based classification [27] and the highest value obtained is 0.85 with logistic regression [26].

$$Precision = \frac{TruePositives}{TruePositives + FalsePositives} \tag{7.1}$$

$$Recall = \frac{TruePositives}{TruePositives + FalseNegatives} \tag{7.2}$$

$$F1 = 2 \cdot \frac{Precision \cdot Recal}{(Precision + Recall)} \tag{7.3}$$

7.3 Data set

7.3.1 Aruba test bed

The open data set is collected in Aruba test bed by monitoring a single elderly woman for eight months. It contains a two-bedroom house with a backyard, a garage, an office and a kitchen. It has the information of 32 wireless motion sensors, 4 temperature sensors and 4 magnetic door sensors. The positions of these sensors can be seen in Figure 7.1. However, this study uses only the binary sensors, i.e. motion and door sensors, to be able to extract patterns from their binary change. On the other hand, as in [30], these non-binary sensors can be converted into binary with a specified for threshold.

7.3.2 Data set and activities

The data set contains two files called data and a README without extensions. It has 1,719,558 lines of recording with 11 distinct activities for 219 days between 2010 and 2011. An example activity is given in Figure 7.2. In addition, the names of the given 11 activities and the total number of occurrences of these activity can be examined in Table 7.2.

The structure of the rows is described in order as follows:

- **Date:** Date of the observation in Year-Month-Day format.
- **Time:** Time of the observation in Hour:Minute:Seconds. Floating Point format.
- **Sensor ID:** Sensor ID of the participated sensor. Sensor IDs classified as 'M' for motion, 'D' for door and 'T' for temperature and the following number indicates the ID of the specified sensor.
- **Sensor Status:** Sensor's state. ON/OFF for motion, OPEN/CLOSED for door and measured temperature in Celsius degrees (floating point number).
- **Annotation:** Annotation of the activity beginnings and endings. No information given, if the line does not specify a beginning or an ending.

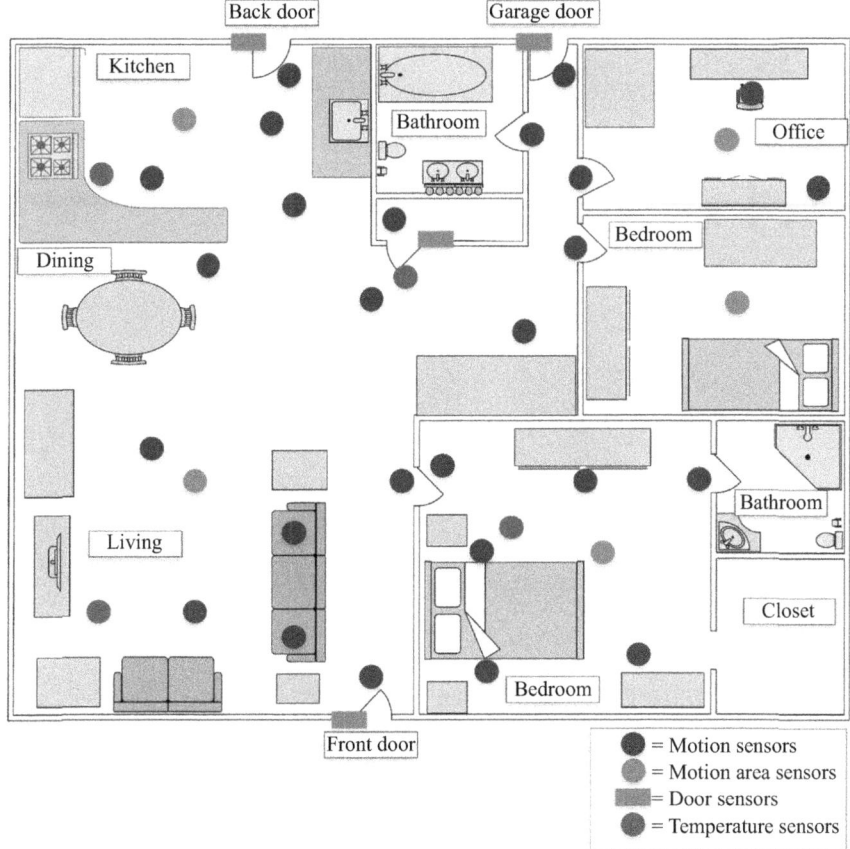

Figure 7.1   House plan of the Aruba data set

7.3.3   Data preparation

To interact with this data set on the programming platforms of R and Python, some main changes have been made on the data set. The main changes discussed in this section are Comma Separated Value (CSV) conversion, missing or corrupted value handling and data set segmentation.

7.3.3.1   CSV conversion

As the data obtained from [1] is a file with no extension like txt or csv, it needed to be converted to a csv file for further processing. To do so, first, annotations are modified since for all the activities have two different action tags called begin and end. These tags are connected to the activity name with an underscore. Second, all the white spaces are converted into commas to have the csv format. Lastly, the newlines of the each row is replaced with commas followed by newline to be able to represent

```
2010-11-04 05:40:51.303739 M004 ON Bed_to_Toilet begin
2010-11-04 05:40:52.342105 M005 OFF
2010-11-04 05:40:57.176409 M007 OFF
2010-11-04 05:40:57.941486 M004 OFF
2010-11-04 05:43:24.021475 M004 ON
2010-11-04 05:43:26.273181 M004 OFF
2010-11-04 05:43:26.345503 M007 ON
2010-11-04 05:43:26.793102 M004 ON
2010-11-04 05:43:27.195347 M007 OFF
2010-11-04 05:43:27.787437 M007 ON
2010-11-04 05:43:29.711796 M005 ON
2010-11-04 05:43:30.279021 M004 OFF Bed to Toilet end
```

Figure 7.2   Bed-to-Toilet activity from raw data set

Table 7.2   Activities and number of activity instances

Class	Activity name	Number of instances
C1	Meal_Preparation	1,606
C2	Relax	2,919
C3	Eating	257
C4	Work	171
C5	Sleeping	401
C6	Wash_Dishes	65
C7	Bed_to_Toilet	157
C8	Enter_Home	431
C9	Leave_Home	431
C10	Housekeeping	33
C11	Resperate	6

NA values in the Annotation column for the activities with no annotation in the csv file.

7.3.3.2   Missing/corrupted value handling

Following the completion of the CSV conversion operation, the data are uploaded to R Studio for further processing. All the rows containing sensor ID containing 'T' letter are eliminated since temperature information is not relevant to the patterns that are created in this study. In the summary of the data set, some corruptions are detected, especially in the columns Sensor ID and Sensor Status. These corruptions and their occurrences are given in Tables 7.3 and 7.4.

To correct the corrupted data when possible, all the values containing 'ON' and 'OPEN' converted into 'ON', the values containing 'OF', 'OFF' and 'CLOSE' converted into 'OFF' and all the other rows with corruption are eliminated from the data set. In this process, only one 'Sleeping' activity is affected. Resultant data set has 1,602,985 lines of only binary sensor data.

Table 7.3   Incorrect values in the data set for Sensor IDs

Corrupted value	Occurrences	Corrupted value	Occurrences
c	1	LEAVEHOME	4

Table 7.4   Incorrect values in the data set for Sensor Status

Corrupted value	Occurrences	Corrupted value	Occurrences
180	1	ON55	1
300	1	ON5c	1
O	1	ONc	44
OcFF	1	ONc5	2
OF	1	ONc5c	2
OFcF	3	ONcc	5
OFF5	14	ONM009	1
OFF5	6	cONM024	1
OFF5cc	2	ONM026	2
OFFc	39	OPENc	1
OFFc5	2	ON5	17
OFFcc	8	OFFccc5	1

For the Sensor IDs, since there is no clue to do error correction on those labels, the rows with corrupted values are eliminated from the data set.

7.3.3.3   Data set segmentation

After handling the problematic data in the data set, an activity-based segmentation has been implemented in two levels with an R script. In the first level, each activity is separated from each other for separate examination. In the 11 distinct activity file stored in the format of "ActivityName.csv" file where ActivityName is the label of the activity. In the second level of segmentation, each instance of an activity is separated and stored in a csv file in the format of "ActivityNameK.csv" file where the k is the kth instance of the specified activity.

During the segmentation process, the sensor events which do not belong to any activity are eliminated.

7.4   Opportunities and directions

Although the data set presented in this chapter seems that it has limited information, various types of information can be extracted using statistical methods. In this section, some of the possible information can be extracted from these kinds of data set are presented.

To show the opportunities that these data sets present as well as their possible effects, five different information is presented. Apart from these information, additional grouping is implemented and examined in this section.

First, to present the interactions between activities and the sensors, a figure for the distribution map of sensor-activity involvement is created as in Figure 7.3. Since our main aim is to detect activities and to understand behaviours of a person, using this kind of information, we can understand which sensor is involved more in which activity. In this figure, grayscale coding is used to show the activeness of each sensor on activities where darker parts mean small or no involvement and lighter part mean high involvement. To illustrate, the sensor 'M002' is only involved with 4 of the 11 activities and its involvement to the activity is lower compared to the other active sensors in those activities. In addition, there are some sensor collections that are used actively with the similar frequencies such as sensors between 'M015' and 'M019' used frequently for 'Meal Preparation' and 'Wash Dishes'. Hence, a robust enough method to distinguish the correct activity in such cases is required. Furthermore, from the figure, it can be observed that the sensor 'D003' has no involvement with the any of the activities and sensor 'M003' involved actively with only one activity. All in all, this figure gives intuitive information about the sensors participated in the pattern of the each activity which may enable researchers or developers to optimize activity recognition models by eliminating non-significant data.

Second, in addition to the involvement of the sensors, sensor event information of the activities is also presented in Figure 7.4. This figure presents the maximum number of events occurred, the mean number of events and the number of instances of those activities in the data set in order. From this figure, it can be said that the activities 'Meal Preparation' and 'Relax' have some outlier cases since their maximum number of events are larger compared to the other activities. On the other hand, considering the average number of sensor events occurred for each activity, 'Housekeeping' is the one with the highest value which indicates that there are more motion in the 'Housekeeping' activity compared to the others.

Third, duration for each activity is also presented in Table 7.5. From Table 7.5, even though some of the activities take similar amount of time, the time spent can be added into classification factors or features for activity prediction. However, it still does not solve the problems like the one between Meal_Preparation and the Wash_Dishes activities which are also described as the activities with lowest prediction accuracy in [13]. On the other hand, it can be seen that mean time spent of each activities can be good feature candidate for learning models.

Another information that can be obtained is the observed probability of each activity, i.e. the probability that an activity occurred in this data set (see Table 7.6). This information may enable researchers to profile the user so that activity prediction model can be further improved. For this specific data set, the main activities performed by the resident are relaxing and meal preparation. In addition, these probabilities and their changes over time while collecting the data can help any system to detect abnormal behaviour. As stated in the Section 7.2, [21] managed to detect urinary tract infections by checking the toilette usage of elderly over time.

	Meal Preparation	Relax	Eating	Work	Sleeping	Wash Dishes	Bed to Toilet	Housekeeping	Resperate	Enter Home	Leave Home	Involved Activities
M001	0.008	0	0	0	0.3	0	0	1.6	0	0	0	5
M002	7E-04	0	0	0	19	0	0	0.1	0	0	0	4
M003	0.005	0	0	0	61	0.019	0	1.1	0	0	0	6
M004	0.062	0	0	0	0.2	0.029	53.53	2.4	0	0	0	8
M005	0.064	0.1	0	0	0.5	0.019	16.77	4.4	0	0	0	8
M006	0.115	0.1	0.1	0	0.7	0.057	0	2.9	0	0.049	0.104	8
M007	0.114	0.2	0.1	0.1	8.9	0.048	29.7	9.9	0	0	0	8
M008	0.286	0.2	0.2	0	0.2	0.115	0	3.9	0	0	0.207	7
M009	1.248	63	1	0	0.1	0.21	0	5.4	0	0	0	6
M010	0.394	6.9	0.4	0	0.1	0.019	0	2.8	0	0	0	6
M011	0.055	0	0.3	0	0.1	0	0	0.6	0	0	0	5
M012	0.418	1.8	0.3	0	0.1	0.076	0	2.2	0	0	0.207	7
M013	0.975	9.1	1.5	0	0.1	0.44	0	5.1	0	0	0.207	8
M014	4.255	1.2	70	0	0.2	3.67	0	6.9	0	0.691	1.087	9
M015	22.05	0.5	2.1	0	0.1	29.6	0	3.4	0	0	0	7
M016	5.796	0.2	0.7	0	0.1	3.881	0	0.8	0	0.099	0.052	8
M017	9.397	0.2	1.6	0	0.1	8.373	0	1.9	0	0	0	6
M018	11.74	0.6	4.4	0	0.2	11.5	0	4.3	0	0.099	0.311	9
M019	33.43	0.8	3.8	0	0.1	36.79	0	5.5	0	0.099	0.207	9
M020	2.815	6.7	6	0.1	0.3	1.95	0	14	0	0.987	0.776	9
M021	1.495	0.6	1.4	0.1	0.2	0.832	0	6.1	0	4.097	1.449	9
M022	0.885	0.5	0.7	0.9	0.2	0.363	0	2.6	1.1	0.987	3.054	10
M023	0.579	0.4	0.4	0.2	0.5	0.057	0	1.2	0	0.197	0.518	9
M024	2.635	4.9	4.5	3.5	5.1	0.898	0	4.9	0	0	0	7
M025	0.069	0.1	0	3.5	0.1	0.134	0	0.2	69	0	0.104	8
M026	0.283	0.5	0	62	0.1	0.249	0	0.9	12	0	0.207	9
M027	0.134	0.2	0	27	0.1	0.153	0	0.5	17	0	0.104	9
M028	0.149	0.1	0.1	2.2	0.1	0.22	0	0.6	1.3	1.974	1.346	10
M029	0.173	0.1	0.2	0.2	0.2	0.182	0	0.7	0	9.921	19.62	9
M030	0.174	0.1	0.1	0.1	0.2	0.076	0	0.2	0	38.55	26.29	9
M031	0.025	0	0	0	0.1	0	0	2.5	0	0	0	4
D001	0.011	0	0.1	0	0	0	0	0.1	0	0	0	5
D002	0.036	0	0	0	0	0	0	0	0	0.197	0.259	7
D003	0	0	0	0	0	0	0	0	0	0	0	0
D004	0.129	0.1	0.1	0.1	0.1	0.038	0	0	0	42.05	43.89	9
Involved Sensors	34	34	30	21	34	28	3	34	5	14	20	

Figure 7.3   Frequency map of involved sensors for each activity

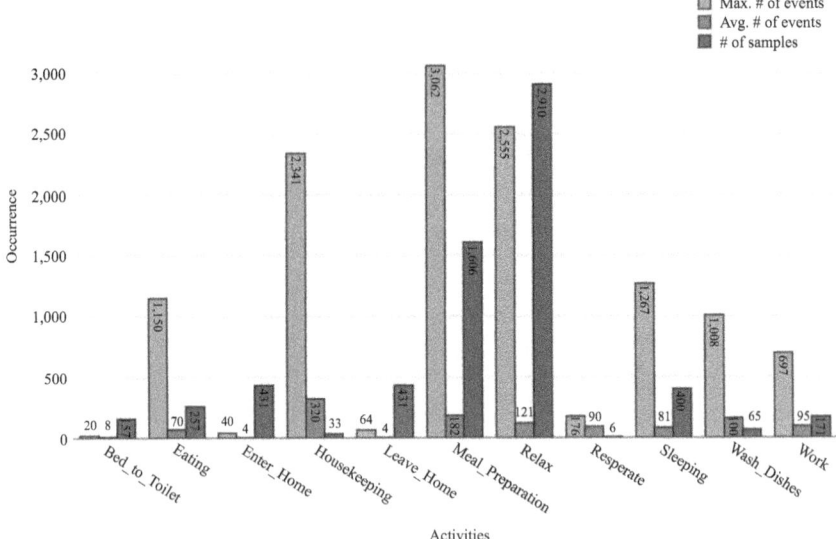

Figure 7.4   Sensor event details of activities

*Table 7.5   Duration information of each activity in **minutes***

Activity	Mean
Meal_Preparation	7.83
Relax	33.45
Eating	10.14
Work	17.08
Sleeping	240.87
Wash_Dishes	7.2
Bed_to_Toilet	2.73
Housekeeping	20.42
Resperate	8.5
Leave_Home	0.020
Enter_Home	0.01

Lastly, the transition of the activities can be examined and organized. To achieve this, different methods can be applied for these data sets. In this chapter, the presented method checks each activities' previous and next activities based on the tags in the annotations such as 'end' and 'begin'. It checks for the first 'begin' tag after an 'end' tag and for the first 'end' tag before 'begin'. The obtained data are converted into probabilities and multiplied by 100 to get the percentages which are presented in Table 7.7.

For simplification, the activities are represented as classes where the details of these abbreviations given in Table 7.2. In Table 7.7, all the values are probabilities

Table 7.6   Observed probabilities of activities

Activity	Observed probability
Bed_to_Toilet	0.03
Eating	0.05
Meal_Preparation	0.29
Relax	0.52
Sleeping	0.07
Wash_Dishes	0.01
Work	0.03

Table 7.7   Activity transition matrix

	C1	C2	C3	C4	C5	C6	C7	C8	C9	C10	C11
C1	37.35	33.58	2.35	2.54	13.11	0.31	0.12	0.06	10.08	0.31	0.19
	37.35	44.71	12	1.61	0.06	0.19	0.06	3.53	0.06	0.31	0.12
C2	24.48	63.09	2.97	1.84	0.65	1.57	0.03	0.03	4.71	0.48	0.14
	18.43	63.11	0.58	2.39	7.61	0.82	0.03	6.18	0.03	0.68	0.14
C3	74.63	7.09	11.94	1.12	0.37	0.75	0.37	0.37	2.61	0.37	0.37
	14.18	33.58	12.31	4.48	1.12	14.18	0.37	18.28	0.37	0.75	0.37
C4	14.29	38.46	6.59	21.43	0.55	3.3	0.55	0.55	11.54	2.2	0.55
	22.53	29.67	1.65	21.43	2.75	1.65	0.55	15.38	0.55	2.75	1.1
C5	0.24	54.26	0.73	1.22	4.14	0.24	37.71	0.24	0.73	0.24	0.24
	51.58	4.38	0.24	0.24	4.14	0.24	37.96	0.49	0.24	0.24	0.24
C6	3.95	31.58	50	3.95	1.32	2.63	1.32	1.32	1.32	1.32	1.32
	6.58	60.53	2.63	7.89	1.32	2.63	1.32	9.21	1.32	5.26	1.32
C7	0.6	0.6	0.6	0.6	92.86	0.6	1.79	0.6	0.6	0.6	0.6
	1.19	0.6	0.6	0.6	92.26	0.6	1.79	0.6	0.6	0.6	0.6
C8	12.44	40.95	11.09	6.33	0.45	1.58	0.23	0.23	23.3	2.94	0.45
	0.23	0.23	0.23	0.23	0.23	0.23	0.23	0.23	97.74	0.23	0.23
C9	0.23	0.23	0.23	0.23	0.23	0.23	0.23	97.74	0.23	0.23	0.23
	36.88	31.22	1.36	4.75	0.68	0.23	0.23	23.53	0.23	0.68	0.23
C10	11.36	45.45	4.55	11.36	2.27	9.09	2.27	2.27	6.82	2.27	2.27
	11.36	31.82	2.27	9.09	2.27	2.27	2.27	29.55	2.27	2.27	4.55
C11	11.76	23.53	5.88	11.76	5.88	5.88	5.88	5.88	5.88	11.76	5.88
	17.65	23.53	5.88	5.88	5.88	5.88	5.88	11.76	5.88	5.88	5.88

of transition from one class to the other in percentage format. The upper values in each cell represents the previous activity transition which means the probability percentage of a class occurring before the selected class, and the lower value is the next transition where it means the probability percentage of a class occurring after the given class. Hence, it expected that the previous and next transition probabilities should be the same for the activity itself (A<–>A), i.e. the upper and lower cells of main diagonal should have same values for each activity. However, it can be observed that these values differ slightly, which indicates that there are problematic cases in the

Table 7.8   Grouping scheme for times of the days

Schema 1	Label
06:00–10:00	Morning
10:00–12:00	Noon
12:00–18:00	Afternoon
18:00–22:00	Evening
22:00–06:00	Night

data set even though a pre-processing and cleaning stage are performed. To illustrate, there is still a slight chance that the resident can leave home even though she has not returned from her latest leaving yet. Nevertheless, this transition table still gives information on which activities could be expected (the ones with higher probability), after a selected activity. It can be observed that C8 and C9 are occurring one after another, so this gives us a semantic information that if a person leaves the house, the next activity is entering the house. Even though it is natural for humans to understand, it is a quite valuable information for any computer algorithm to make robust activity classification.

All the methods mentioned so far have obtained information from a cleaned data set without modification. However, to understand more about the resident and its behaviour, it is possible that for minor additions can be done. For this purpose, we have presented grouping based on period of the days such as morning and noon. In this chapter, only one segmentation schema is given in Table 7.8, but many others can be implemented as well.

Having such groups, we can examine resident's habits more closely by checking each period individually. To illustrate, bar charts of activities are presented for each period of the day in Figure 7.6. This figure shows us that some of the activities are performed only in certain parts of the days. For example, bed-to-toilet activity is performed mainly in night period. Similarly, we may distinguish meal preparation and wash dishes activities by using their distributions over time periods as the resident is more likely to prepare a meal in the morning and evenings than afternoon which is not the case for washing dishes.

Similar to Figure 7.5, activity lengths by daytimes can also be examined to understand the time spent for those activities in different periods as presented in Figure 7.7.

All in all, this section presented various methods to obtain information that may enhance activity recognition, resident profiling and abnormality detection.

In our next study, multinominal logistic regression will be used to model nominal outcome variables, in which the log odds of the outcomes are modelled as a linear combination of the explanatory variables. Probabilities of activities will be calculated based on different time periods.

The ratio of the probability of choosing one outcome category over the probability of choosing the baseline category is often referred as relative risk (and it is sometimes referred to as odds). The relative risk is the right-hand side linear equation

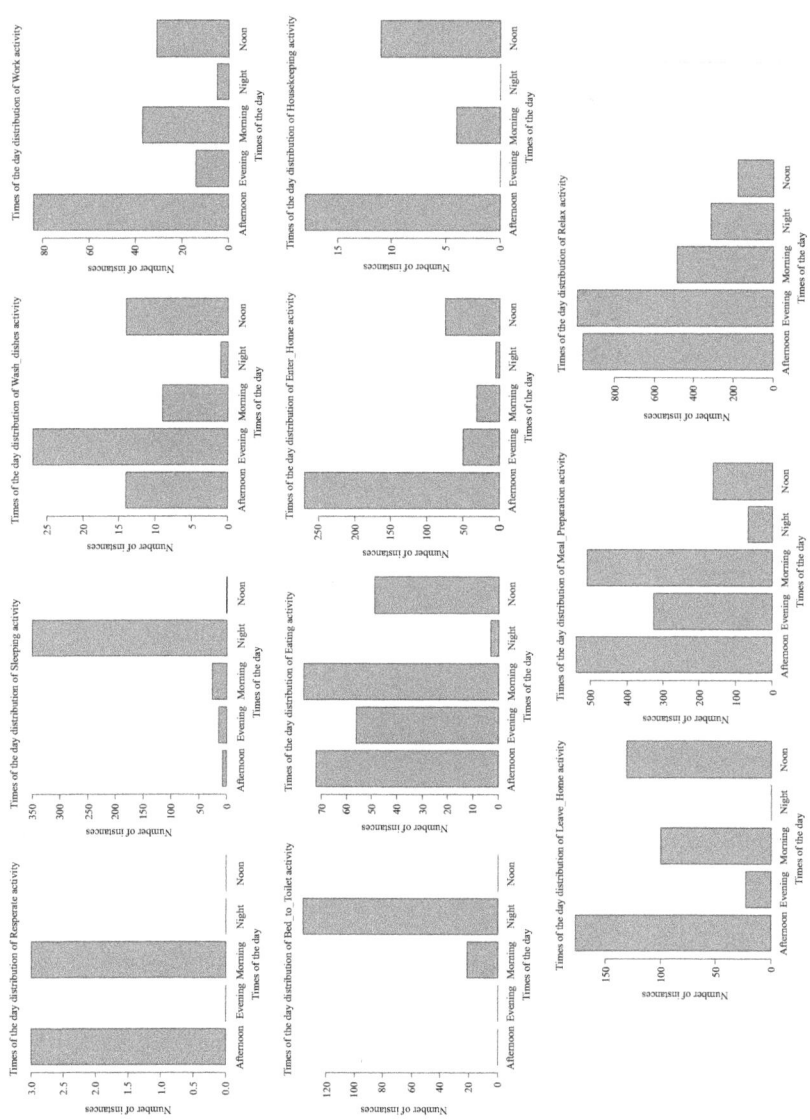

Figure 7.5   Bar charts of activity instances for groups in Schema 1

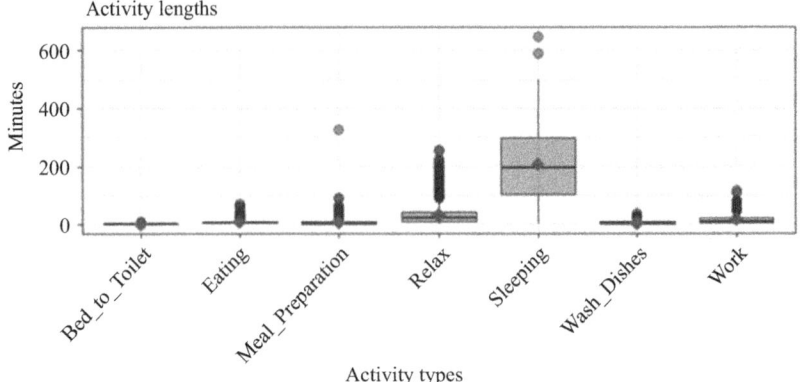

Figure 7.6   Activity lengths by types

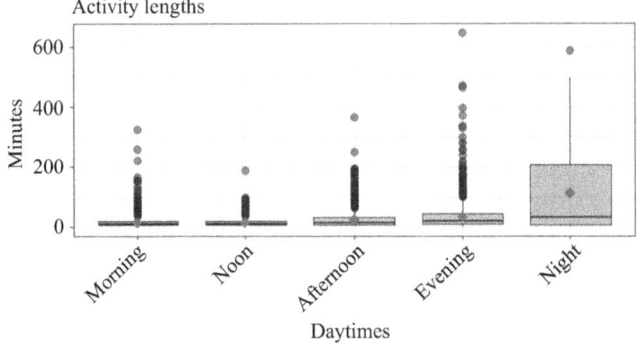

Figure 7.7   Activity lengths by daytimes

exponentiated, leading to the fact that the exponentiated regression coefficients are relative risk ratios for a unit change in the predictor variable.

It is also required to perform analysis of the movements of the elderly person in the house in which the data set is collected. These movements show the time spent in different home activities, such as bed-to-toilet, meal preparation and eating, wash-dishing, resting, working and sleeping. On this type of multinominal logistic modelling, it is also possible to take into account the time during the day activities take place together with all other available explanatory variables in the data set. The results of such an analysis have good potential to further shed light on the pattern of behaviours of people having very different background and patient characteristics.

7.5   Conclusion

During the last decade, we have witnessed the emergence of various application areas of IoT. This popularity allowed researchers to gather or to reach to data sets in different areas. One of these recently emerged areas called E-Health become more visible in the literature with its variety of public data sets. Hence, the analysis of these data sets is required to understand and to extract higher level information.

In this chapter, we focused on understanding ADL from the unobtrusive binary sensors deployed in a smart home. Even though there are many applications that does ADL recognition with numerous methods, understanding the data well is essential to provide higher accuracy and to have better control on the system. To emphasize these, we select an exemplary data set to present the possible problems can be faced during the pre-processing stage. Lastly, we present a statistical method along with an additional feature extracted from the time information of the data, i.e. period of the day that can be used either for classification or for enhancing the existing models for ADL recognition along with the others given in the Section 7.2 such as CRF and DBN.

References

[1] Cook D. Learning setting-generalized activity models for smart spaces. IEEE Intelligent Systems. 2012;27(1):32–38.

[2] Cook DJ, Crandall AS, Thomas BL, *et al.* CASAS: A smart home in a box. Computer. 2013;46(7):62–69.

[3] Lin J, Yu W, Zhang N, *et al.* A survey on Internet of Things: Architecture, enabling technologies, security and privacy, and applications. IEEE Internet of Things Journal. 2017;4(5):1125–1142.

[4] HKExnews. Size of the global Internet of Things (IoT) market from 2009 to 2019 (in billion U.S. dollars). Available from: https://www.statista.com/ statistics/485136/global-internet-of-things-market-size/.

[5] Peterson C. From innovation to implementation: eHealth in the WHO European region. Copenhagen, Denmark: WHO Regional Office for Europe, 2016.

[6] Guest editorial: Smart cities and smart sensory platforms. IET Wireless Sensor Systems. 2018;8(6):247–248.

[7] Al-Turjman F, Mostarda L, Ever E, *et al.* Network experience scheduling and routing approach for big data transmission in the Internet of Things. IEEE Access. 2019;7:14501–14512.

[8] Pino T, Choudhury S, and Al-Turjman F. Dominating set algorithms for wireless sensor networks survivability. IEEE Access. 2018;6:17527–17532.

[9] Al-Turjman F, Imran M, and Vasilakos A. Value-based caching in information-centric wireless body area networks. Sensors. 2017;17(12):181. Available from: http://dx.doi.org/10.3390/s1701018

[10] Monekosso D, Florez-Revuelta F, and Remagnino P. Ambient Assisted Living [Guest editors' introduction]. IEEE Intelligent Systems. 2015;30(4):2–6.

[11] Chen D, Bharucha AJ, and Wactlar HD. Intelligent video monitoring to improve safety of older persons. In 2007 29th Annual International Conference of the IEEE Engineering in Medicine and Biology Society, Aug 2007; 3814–3817.

[12] Papakostas M, Giannakopoulos T, Makedon F, *et al.* Short-term recognition of human activities using convolutional neural networks. In: 2016 12th International Conference on Signal-image Technology Internet-based Systems (SITIS); 2016. pp. 302–307.

[13] Gochoo M, Tan T-H, Liu S-H, *et al.* Unobtrusive activity recognition of elderly people living alone using anonymous binary sensors and DCNN. IEEE Journal of Biomedical and Health Informatics. 2018;23(2):693–702.

[14] Abbate S, Avvenuti M, and Light J. MIMS: A minimally invasive monitoring sensor platform. IEEE Sensors Journal. 2012;12(3):677–684.

[15] Ha S, Yun J, and Choi S. Multi-modal convolutional neural networks for activity recognition. In: 2015 IEEE International Conference on Systems, Man, and Cybernetics; 2015. pp. 3017–3022.

[16] Akyildiz IF, Melodia T, and Chowdhury KR. A survey on wireless multimedia sensor networks. Comput Netw. 2007;51(4):921–960. Available from: http://dx.doi.org/10.1016/j.comnet.2006.10.002.

[17] CASAS W. Dataset. Available from: http://casas.wsu.edu/datasets/.

[18] van Kasteren T, Noulas A, Englebienne G, *et al.* Accurate activity recognition in a home setting. In: Proceedings of the 10th International Conference on Ubiquitous Computing, ser. UbiComp 08. New York, NY, USA: Association for Computing Machinery. 2008; p.19. Available from: https://doi.org/10.1145/1409635.1409637

[19] Michael J, Grießer A, Strobl T, *et al.* Cognitive modeling and support for ambient assistance. In: Lecture Notes in Business Information Processing. Springer: Berlin, Heidelberg; 2013. pp. 96–107.

[20] Sprint G, Cook DJ, Fritz R, *et al.* Detecting health and behavior change by analyzing smart home sensor data. 2016 IEEE International Conference on Smart Computing (SMARTCOMP); 2016. pp. 1–3.

[21] Fritz RL, and Cook D. Identifying varying health states in smart home sensor data: An expert-guided approach. In World Multi-Conference of Systemics, Cybernetics and Informatics: WMSCI 2017, 2017.

[22] Rantz MJ, Skubic M, Koopman RJ, *et al.* Using sensor networks to detect urinary tract infections in older adults. In: 2011 IEEE 13th International Conference on e-Health Networking, Applications and Services; 2011. pp. 142–149.

[23] Krishnan NC, and Cook DJ. Activity recognition on streaming sensor data. Pervasive and Mobile Computing. 2014;10:138–154. Available from: http://www.sciencedirect.com/science/article/pii/S1574119212000776.

[24] Cook DJ, Krishnan NC, and Rashidi P. Activity discovery and activity recognition: A new partnership. IEEE Transactions on Cybernetics. 2013;43(3):820–828.

[25] Xu L, Wang G, and Guo X. A two-layer framework for activity recognition with multi-factor activity pheromone matrix. In MATEC Web of Conferences. EDP Sciences, 2018;189:10001.

[26] Wan J, Li M, O'Grady M, *et al.* Time-bounded activity recognition for ambient assisted living. IEEE Transactions on Emerging Topics in Computing. 2018; pp. 1–1.

[27]   De Paola A, Ferraro P, Gaglio S, *et al*. Context-awareness for multi-sensor data fusion in smart environments. In: Adorni G, Cagnoni S, Gori M, *et al*., editors. AI*IA 2016 Advances in Artificial Intelligence. Cham: Springer International Publishing; 2016. pp. 377–391.

[28]   Machot FA, and Mayr HC. Improving human activity recognition by smart windowing and spatio-temporal feature analysis. In: Proceedings of the 9th ACM International Conference on Pervasive Technologies Related to Assistive Environments, ser. PETRA 16. New York, NY, USA: Association for Computing Machinery, 2016. Available from: https://doi.org/10.1145/2910674.2910697

[29]   Agarwal M, and Flach P. Activity recognition using conditional random field. In: Proceedings of the 2nd International Workshop on Sensor-Based Activity Recognition and Interaction, ser. iWOAR 15. New York, NY, USA: Association for Computing Machinery, 2015. Available from: https://doi.org/10.1145/2790044.2790045

[30]   Kröse B, Van Kasteren T, Gibson C, *et al*. Care: Context awareness in residences for elderly. In: International Conference of the International Society for Gerontechnology, Pisa, Tuscany, Italy, 2008, pp. 101–105.

Chapter 8

Cybersecurity attacks on medical IoT devices for smart city healthcare services

Marina Karageorgou[1], Georgios Mantas[2,3], Ismael Essop[3], Jonathan Rodriguez[2,4] and Dimitrios Lymberopoulos[1]

Smart city is an emerging concept whose main goal is to improve the quality of life of its citizens by leveraging Information and Communications Technologies (ICTs) as the key medium. In this context, smart city healthcare can play a pivotal role toward the improvement of citizens' quality of life, since it can allow citizens to be provided with personalized e-health services, without limitations on time and location. In smart city healthcare, medical Internet of Things (IoT) devices constitute a key underlying technology for providing personalized e-health services to smart city patients. However, despite the significant advantages that IoT medical device technology brings into smart city healthcare, medical IoT devices are vulnerable to various types of cybersecurity threats and thus, they pose a significant risk to smart city patient safety. Based on that and the fact that the security is a critical factor for the success of smart city healthcare services, novel security mechanisms against cyberattacks of today and tomorrow on IoT medical devices are required. Toward this direction, the first step is the comprehensive understanding of the existing cybersecurity attacks on IoT medical devices. Thus, in this chapter, we will provide a categorization of cybersecurity attacks on medical IoT devices which have been seen in the wild and can cause security issues and challenges in smart city healthcare services. Moreover, we will present security mechanisms, derived from the literature, for the most common attacks, as well as highlight emerging good practice and approaches that manufacturers can take to improve medical IoT device security throughout its life cycle. In this chapter, the authors' intent is to provide a foundation for organizing research efforts toward the development of the proper security mechanism against cyberattacks targeting IoT medical devices.

8.1 Introduction

The concept of smart city in a few years will be a global reality. In order to better understand the smart city, we must realize that it concerns all the areas of our everyday

[1]Department of Electrical and Computer Engineering, University of Patras, Patras, Greece
[2]Mobile Systems Group, Instituto de Telecomunicações, Aveiro, Portugal
[3]Faculty of Engineering and Science, University of Greenwich, London, UK
[4]Faculty of Computing, Engineering and Science, University of South Wales, Pontypridd, UK

life, such as work, transport, and healthcare. The concept of the smart city has not been strictly defined, but a definition of a smart city might be given as follows: "A Smart City is a designation given to a city that incorporates information and communication technologies (ICT) to enhance the quality and performance of urban services such as energy, transportation and utilities in order to reduce resource consumption, wastage and overall costs. The overarching aim of a Smart City is to enhance the quality of living for its citizens through smart technology" [1].

In this context, smart city healthcare can play a significant role toward the improvement of citizens' quality of life, since it can allow citizens to be provided with personalized e-health services without limitations on time and location. In smart city healthcare, medical IoT devices constitute a key underlying technology for providing personalized e-health services to smart city patients. Nevertheless, despite the significant advantages that medical IoT device technology brings into smart city healthcare, medical IoT devices are susceptible to various types of cybersecurity threats and thus, they pose a significant risk to smart city patient safety. For instance, adversaries can compromise hospital systems via medical IoT devices or can hack into medical IoT devices themselves and modify prescriptions or manipulate device's functionality. Based on that and the fact that the security is critical factor for the success of smart city healthcare services, novel security mechanisms against known and unknown cyberattacks on medical IoT devices are required. To this end, the first step is the comprehensive understanding of the existing cybersecurity attacks on medical IoT devices. Therefore, in this chapter, we will provide a categorization of cybersecurity attacks on medical IoT devices which have been seen in the wild and can cause security issues and challenges in smart city healthcare services. Besides that, we will present security mechanisms, derived from the literature, for the most common attacks, as well as highlight emerging good practice and approaches that manufacturers can take to improve medical IoT device security throughout its life cycle. In principle, the authors' intent is to provide a foundation for organizing research efforts toward the development of the proper security mechanisms against cyberattacks targeting medical IoT devices.

Following the introduction, this chapter is organized as follows. In Section 8.2, we give an overview of the smart city healthcare concept and examples of healthcare systems that fit suitably into this concept. In Section 8.3, we provide an overview of medical IoT devices for smart city healthcare systems and present their main security objectives. In Section 8.4, a categorization of cybersecurity attacks on medical IoT devices is presented. In Section 8.5, security countermeasures, derived from the literature, for current existing attacks are discussed. In Section 8.6, emerging good practice and approaches for manufacturers are provided. Finally, Section 8.7 concludes the chapter.

8.2   Smart city healthcare

8.2.1   Overview

Thanks to advances in ICT, there are various networked sensor-based systems and devices that have been developed even on a scale of cities that facilitate and support everyday life. Such smart environments really lead to smart cities that can support their residents' activities to improve quality of life and ensure viability in many areas such as healthcare, transport, entertainment, and professional and social life. The IoT is the basic technology for building smart cities as they enable real-time communication and collaboration between entities/objects and the exchange of information among users. Recently, IoT technologies have made their entry into the wider healthcare sector in a smart city environment [2]. The idea of medical care within the smart city has already begun to be implemented through e-health. In particular, electronic health record (EHR), e-prescription, and medical information systems are precursors to this new concept of smart city healthcare that takes advantage of the benefits of IoT.

8.2.2   Smart city healthcare systems

Various smart city healthcare systems are available, which are either already in use today or are planned and will work in the coming years [3]. Some of these are described in the following.

8.2.2.1   Telemonitoring systems

In this type of systems, an elderly patient or resident who lives alone is being monitored (e.g., vital signs) through an interconnected device (e.g., a smartphone) during the day. When anomalies are detected, alarms are sent to stakeholders who may be related with health profession, social interaction services, etc [4,5]. Telemonitoring systems can help improve the quality of life of either an elderly person or a person with motor or mental problems by helping them to become independent and be an active member of a smart city.

8.2.2.2   Smart drug delivery

Information technology, cloud connectivity, and smart phone access are penetrating the fields of drug delivery devices. Such development opens up a new phase of device innovation that looks at interfaces beyond the device [6]. An example of a consumer-friendly device is that of "smart" drug dose management. In particular, it is a medical IoT device which adjusts the dose of the drug based on the patient's daily, weekly, or monthly needs for treatment intake. All required steps related to needle insertion, injection, end of injection feedback, and needle safety steps are performed automatically as in case of insulin pump (see Figure 8.1) [7].

8.2.2.3   Remote healthcare provision system

A remote healthcare provision makes it possible to offer medical services remotely. The solution does not replace traditional healthcare but it is complementary to it by offering an extended range of services without the need of additional medical staff. The implementation of this kind of systems improves efficiency and thus more patients can be handled at the same time. Using telecommunication tools that offer continuous health monitoring, as well as prophylactic and control tests at home, it improves the efficiency of treatment and provides patients with a greater sense of safety by assuring permanent contact with qualified medical staff. Apart from quick

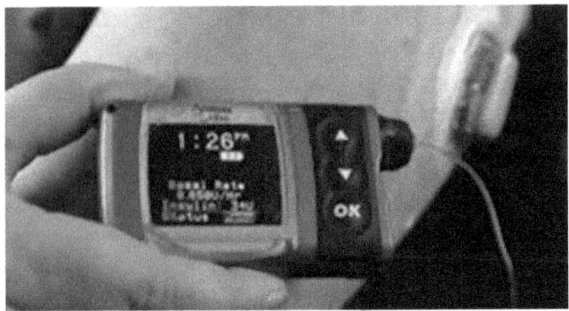

Figure 8.1   Insulin pump case [5]

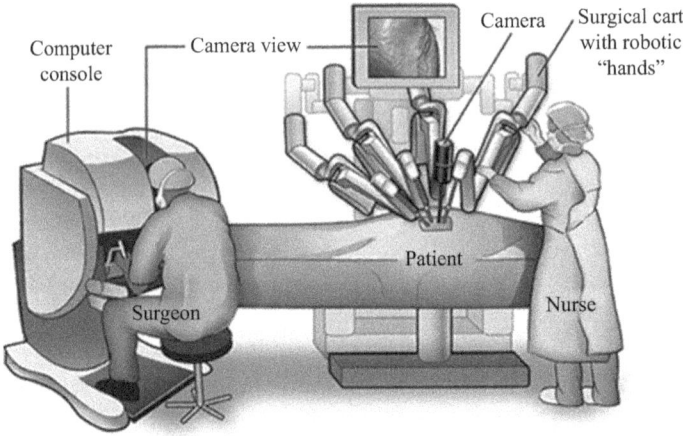

Figure 8.2   Robotic cardiac surgery [5]

and secure access to electronic medical documentation, doctors can also offer tele-consultations to patients at any time and place which leads to savings in terms of time and money [8].

8.2.2.4   Smart surgery

Generally, a computer-assisted surgery system consists of "smart" tools that have an attached controller which, through sensors, images, etc., either gives information about the anatomy of the area to the surgeon or adjusts the tools automatically to the environment. The following classes of robotic-derived surgical devices/systems are considered: (1) handheld tools augmenting the capabilities of the surgeon; (2) tele-operated surgical tools; and (3) autonomous surgical robots (see Figure 8.2) [9].

8.2.3 Benefits arising from smart city healthcare

The benefits of adopting the concept of smart city healthcare can be summarized as follows:

- *Reduce human errors*: Through the use of interconnected medical devices and automated "smart" networks and systems, it is possible to reduce the participation of the human factor, thus reducing the likelihood of medical error.
- *Availability and accessibility*: Patients and caregivers can have access to smart city healthcare services and data without limitations in terms of time and place [10].
- *Patient-centric healthcare system*: The provision of medical care is now completely individualized according to the characteristics and needs of each patient. This means a better and faster response to the treatment and rehabilitation of both short- and long-term illnesses.
- *Ease of use*: Users can easily adopt the use of such systems, as they are either wearable, so they only need to be turned on and off in one click, or they are fully automated [10].
- *Lifetime recording*: Patients can access comprehensive additive data about their past, present, and future health [10].
- *Larger physician involvement*: With the advancement of technology, doctors have access to more information about patients in a more effective and efficient manner. [10].

Finally, it is worthwhile to mention that the adoption of more and more technologies aimed at improving people's health and lifestyles improves the level of health provision and opens new horizons in addressing diseases, peculiarities, and crises at individual and community level.

8.3 IoT medical devices for smart city healthcare systems

8.3.1 Overview

In [11], WHO defines a medical device as any apparatus, appliance, material, or other article—whether used alone or in combination, including the software intended by its manufacturer to be used specifically for diagnostic and/or therapeutic purposes and necessary for its proper application—intended by the manufacturer to be used for human beings for the purpose of:

- diagnosis, prevention, monitoring, treatment, or alleviation of disease;
- diagnosis, monitoring, treatment, alleviation of, or compensation for an injury;
- investigation, replacement, modification, or support of the anatomy or of a physiological process;
- supporting or sustaining life;
- control of conception;
- disinfection of medical devices;
- providing information by means of in vitro examination of specimens derived from the human body;

and does not achieve its principal intended action by pharmacological, immunological, or metabolic means, but which may be assisted in its function by such means [11].

Over the past few years, there is an increasing trend of manufacturing medical IoT devices that are enabled to connect to personal computers, mobile devices, or even to remote servers in hospitals via the Internet. This emerging interconnected world has created new and unique opportunities to improve the provided healthcare services to the citizens of the upcoming smart cities [12]. In principle, all devices fall into four basic categories [13]:

- noninvasive devices;
- invasive medical devices;
- active medical devices;
- special Rules (including contraceptive, disinfectant, and radiological diagnostic medical devices).

The US Food and Drug Administration (FDA) provides another way of classifying medical devices based on the potential risk of harm. They are distinguished in Classes I, II, and III from the minimum to maximum risk. Devices, such as language hatches and handheld surgical instruments, belong to Class I and are subject to minimum regulatory control, general controls. Higher-risk medical devices, such as infusion pumps and surgical needles, are classified in Class II, where they are subject to specific controls, including special labeling, mandatory performance standards, and post-sale monitoring. Class III devices, such as cardiac pacemakers and neurostimulators, are more invasive and pose a much more significant risk, against which neither general nor special control is sufficient to assure safety and effectiveness. Such a device requires premarket approval, in addition to the general controls [14,15].

8.3.2   Medical IoT devices security objectives

The ever-growing adoption of technology does not allow us to secure the interconnected medical devices. It is possible that some of the devices and systems connected to the Internet may expose unintentionally information about us and our surroundings online, and could potentially threaten everyone's safety and security [16].

Safety is a general term that refers to the design, implementation, operation, and maintenance of systems and devices to avoid the occurrence of damaging situations that may lead to injury or loss of life or unintentional environmental damage [17]. On the other hand, security refers to the preservation of privacy of health information, and precisely to the means used in order to achieve that and to assist the caregivers to maintain that information confidential [18]. As we understand, the need to keep medical IoT devices secure becomes necessary, so there should be specific security objectives that we should meet. According to [19], the specific security requirements may fall into the following categories: information; access level; and functional.

8.3.2.1   Information security objectives

This category includes confidentiality, integrity, nonrepudiation, and freshness.

- **Confidentiality**: It assures that confidential information or data are not available to unauthorized users. When medical IoT devices are interconnected to a network and transmit data, there are concerns about the exploitation of the transmitted medical data that can easily be targeted by cyberattackers, who may exploit them for their own benefit, e.g., money and influence on public opinion.

- **Integrity**: It addresses the unauthorized alteration of data [20]. In other words, integrity guarantees that data stored or in transit have not been changed [19]. An attacker may modify the data and compromise the integrity of a medical device. For example, in an insulin pump, where there is a predetermined dose according to the needs of the organism, if the integrity of the device is compromised and the data are changed by an attacker, then the patient's health will be at risk.
- **Nonrepudiation**: It prevents an entity from denying previous commitments or actions [20]. For instance, a medical IoT device should not deny sending a message that was previously sent.
- **Freshness**: It ensures that the messages sent are recent and no old messages have been replayed [19]. Thus, it prevents replay attacks that can be very dangerous to patient health, especially if the interconnected device is to administer a drug dose.

8.3.2.2 Access-level security objectives

The access-level category includes authentication and authorization.

- **Authentication**: It is related to identification and applies to both entities and transmitted information itself. Two entities entering into a communication should identify initially each other. This type of authentication is referred to as entity authentication. On the other hand, transmitted information should be authenticated in terms of its origin and this type of authentication is called origin authentication [19,20]. For example, if authentication is not provided, a malicious actor could interfere between the sender (i.e., medical IoT device) and the receiver (e.g., an IoT gateway and a remote server) in order to intercept medical data or modify device operation.
- **Authorization**: It ensures that only the authorized devices and the users obtain access to the network services or resources [19].

8.3.2.3 Functional security objectives

This category includes availability.

- **Availability**: It ensures that medical IoT devices work properly and there is no denial of service to legitimate entities. That is to say, availability enables a medical IoT device to be used anytime and anywhere, when needed, despite denial-of-service (DoS) attacks or the presence of failures [19].

8.4 Cybersecurity attacks on medical IoT devices

In this section, we provide a review of cyberattacks on medical IoT devices in terms of (a) the attack vectors that they use to compromise the device, (b) the actions performed once the device is compromised, and (c) the impact on the targeted device [16,17,21].

8.4.1 Attack vectors

Nowadays, there are a variety of attack vectors that can be used to compromise medical IoT devices. The most predominant attack vectors include malware, exploitation of software vulnerabilities, and the wireless connectivity.

- **Malware**: It is malicious code inserted into the device, usually stealthy, in order to compromise the confidentiality, integrity, or availability of the device or disrupting the device itself [16,17]. The most common types of malware targeting medical IoT devices are the following:
- **Virus**: A program that is run unwittingly by the user, executes on the victim device, and spreads to other executable programs [17]. Due to a virus running on the device, the programmed execution of certain functions by the device, such as the frequency adjustment of a pacemaker, may be blocked or modified with fatal consequences for the patient.
- **Rootkit**: A software package installed on a device to maintain hidden access to that device with administrator or root privileges. With root access, the attacker is going to have complete control of the devices [17].
- **Trojan**: Any malicious program that is concealed and run by the user [17]. When it is invoked, then it performs some unwanted or harmful function. In particular, in a medical IoT device, the patient could "accidentally" run a Trojan that is presented as the device's operating program or a software update (patch).
- **Worm**: A malware program that spreads itself (i.e., without user intervention) to infect other devices [17]. On a medical IoT device, a Worm could be pre-installed and be activated simply by starting the device or by activating device's connection to a network.

Exploitation of software/firmware vulnerabilities: This is related to intentional use of known weaknesses in device's software or embedded firmware. Firmware updates are usually done separately from the device's software, which requires patients to visit the healthcare professional in charge of implementing the proper changes. In a noticeable example in August 2017, the US FDA recalled 500.000 pacemakers because of firmware vulnerabilities that could allow attackers to access the device remotely and manipulate pacing and battery strength. Moreover, in 2015, a vulnerability that allowed unauthenticated user's root access was discovered in specific type of infusion pump. That was because the device did not verify the authenticity of the firmware updates before installing them. At this point, we have to mention that the attacks exploiting firmware vulnerabilities can allow attackers to modify device's functionality significantly, and thus putting the health of patients at risk [16].

Wireless connectivity: The wireless connectivity of the device leads to a susceptible environment for the attacker to infiltrate. The attacker can obtain access to the device through a tiny radio transmitter that allows adjustment to any settings functions. For instance, the attacker will be enabled to adjust settings on an infusion pump to release large doses of insulin into the patient (i.e., victim) without his/her knowledge or consent (REF Cybersecurity Threats Targeting Networked Critical Medical Devices) [22].

8.4.2 Post-compromise actions

In this section, we review that the different types of actions may be carried out after the compromise of a medical IoT device.

Denial of control action: Denial of control action means that the device operation is disrupted by delaying or blocking the flow of information, denying device availability or networks used to control the device. This can be achieved by installed malware depleting device's resources and is considered as DoS attack against the device [17].

Device, application, configuration, or software manipulation: This means that device, software, or configuration settings are maliciously modified so that the medical IoT device produces unpredictable results [17]. This type of post-compromise actions can be achieved by privilege escalation that exploits bugs, design flaws, or configuration oversights on the device to allow attackers to gain elevated access to resources that they are not authorized for. For example, an attacker could acquire rights to manage a medical device, by exploiting system vulnerabilities and managing its entire function, accessing data as well as altering it. Privilege escalation occurs in two forms: (a) vertical privilege escalation where a lower privilege user or application accesses functions or content reserved for higher privilege users or applications (e.g., the password for a medical device's initialization can be bypassed); (b) horizontal privilege escalation where a normal user accesses functions or content reserved for other normal users [23].

Spoofed device information: False information is sent to operators either to disguise unauthorized changes or to initiate inappropriate actions. Medical devices can be spoofed by tracking and stamping their data while they are being transported, or perhaps sending out incorrect data [24].

Device functionality manipulation: When a device's functionality is maliciously manipulated, unauthorized changes are made to embedded software, alarm thresholds are changed, or unauthorized commands are issued to devices, which could potentially result in damage to equipment (if tolerances are exceeded), premature shutdown of devices and functions, or even disabling medical equipment [17]. In addition, when device functionality manipulation occurs then impersonation attacks can be performed by the compromised device. Therefore, the device can pretend that is another device for the purpose of fraud [25]. Moreover, when device functionality manipulation occurs then replay attacks can be performed by the device as well. A replay attack can occur when an attacker (e.g., through a compromised device) intercepts and saves old messages, which were earlier sent by a legitimate device, and then tries to send them later, impersonating the legitimate device. This type of attack can be part of a Man-in-the-middle (MitM) attack in a network of medical IoT devices [26].

Safety functionality modified: Safety-related functionalities are maliciously manipulated so that they do not operate when needed, or perform incorrect control actions, potentially leading to patient harm or damage to medical equipment [17].

8.4.3   Impact

Cyberattacks could be classified, according to their impact, into three main categories: attacks that cause physical destruction of a medical IoT device, attacks targeting privacy compromise, and attacks targeting patient's health [27].

Physical destruction of a medical device: Such attacks are aimed at physically destroying a medical device. This can happen for competitive gains mainly from rival companies, financial compensation, etc. The physical destruction of an IoT medical device can be done by an attacker either by invading the device directly or by taking advantage of a possible point of entries from the entire network it connects to. Also, this destruction can be done in various ways such as those described above. For example, malware could be used to make unauthorized changes to embedded software [17].

Privacy compromise: An attacker who has accessed data from a medical IoT device compromises privacy as sensitive information is in the hands of an unauthorized person who can use it either for his/her own benefit or simply to harm the patient. For example, by eavesdropping transmitted messages, an attacker may determine whether a person carries an implant or not and if so, then the attacker may find out the type of the implant, its model, serial number, etc. In addition, by capturing telemetry data an attacker may disclose private information about the patient such as the ID of his/her health records, name, and age. In all cases, the overall result is a serious compromise of patient's privacy [28].

Patient's health: Attacks aimed at exploiting or reversing the operation of medical IoT devices can cause malfunctions in patients' health or even death. For example, if a pacemaker does not operate at the preset frequencies, then it is very likely that the patient's health is exposed irreparably. Another example is the insulin pump, which injects certain amounts of insulin at specific intervals to the patient. If an attack is aimed at altering its function, then the life of the patient will be compromised [28]. The potential adverse events of a cybersecurity attack on a connected implantable medical device (IMD) on the patient's health are presented in Figure 8.3.

8.4.4   Impact evaluation

Cyberattacks on medical devices can have a huge impact on the lives of many people in various ways. In particular, they may cause damage to physical equipment (i.e., medical IoT devices), compromise patients' privacy (e.g., data leakage, identity theft), or put at risk patients' health (e.g., adverse events). Furthermore, cyberattacks on medical devices may have a negative financial impact on the healthcare industry. The healthcare sector is an industry whose goal is to provide high quality of healthcare services to patients. If this provision is hampered by cyberattacks, the financial damage to both healthcare providers (e.g., public/private hospitals) and medical devices manufacturers is likely to become enormous. Moreover, the impact of cyberattacks on

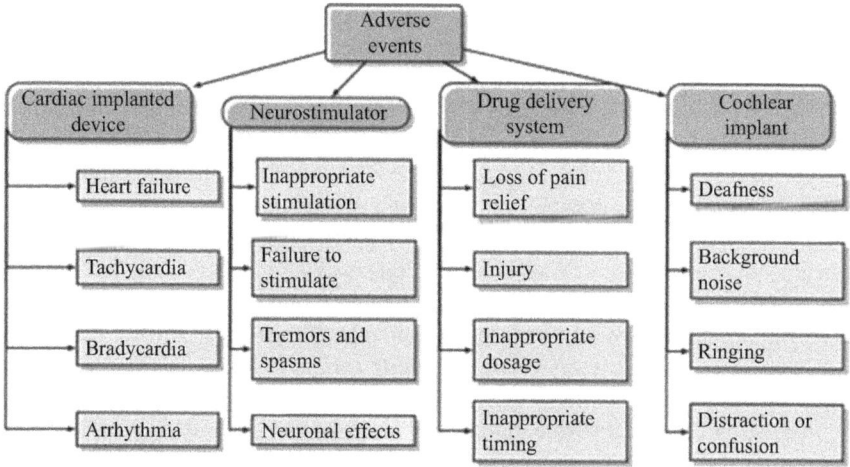

Figure 8.3   Adverse events for four representative IMD types (Source: [28])

interconnected medical IoT devices may be reputational as well. The impact it may have is a damaged public perception of the device's quality which may negatively affect supplier/manufacturer's relationships with their customers and reduce business opportunities. On the other hand, cyberattacks on medical IoT devices may also have a psychological impact on patients, as cyberattacks' consequences may cause confusion, frustration, or anxiety to patients who may think they did not follow the proper procedures and their life is probably at risk. Finally, the harm of a cyberattack on an IoT medical device may also have social extensions. Particularly, it may cause disruption in daily life activities and probably negative changes in public perception of healthcare or even technology.

8.5   Security countermeasures

In this section, we present countermeasures (i.e., security mechanisms) that could be adopted against the different cyberattacks on medical IoT devices discussed in Section 8.4. These security mechanisms are to ensure the security objectives of medical IoT devices identified in Section 8.3. Thus, we have considered security mechanisms for ensuring (i) confidentiality and integrity; (ii) availability; (iii) authentication and authorization; and (iv) nonrepudiation.

8.5.1   Ensuring confidentiality and integrity

Data stored on medical IoT devices or in transit should be kept confidential. In other words, data should not be made available or disclosed to unauthorized entities and this can be achieved through data encryption where symmetric key ciphers (i.e., block and stream ciphers) and public-key ciphers are used. However, the high-power consumption and implementation costs of public-key ciphers preclude their use for data encryption on the resource-constrained medical IoT devices. Hence, symmetric key ciphers constitute the only practical option for data encryption on this kind of devices [14]. Nevertheless, symmetric key ciphers suffer from the key distribution problem where the suitability of a particular key distribution scheme depends on (a) the type of the medical IoT devices, (b) the expected interactions with other devices (e.g., smartphone/body gateway), and (c) other assumptions relevant to the operational environment [28]. For instance, fixed pre-configured keys in the medical IoT devices are vulnerable to compromise. Moreover, secret keys should be updated automatically, since many users (e.g., elderly people) are unable or unwilling to configure secure secret keys or update them frequently. In principle, shared keys should be generated with high agreement between the two communicating entities, high randomness, at a first rate, and with a minimum computational/energy overhead [14]. Consequently, the generation of shared keys comprises a challenge for medical IoT devices and a number of works have already been proposed to deal with it.

On the other hand, data integrity ensures that stored and in-transit data have not been altered by unauthorized or unknown means. That is to say that data stored in the medical IoT devices or being transmitted through wireless links can only be modified by authorized entities. For instance, if there is no integrity-checking mechanism on an IMD (e.g., pacemakers, defibrillators), data can be modified, during the transmission over insecure radio channels, without the modification being detected. Moreover, the lack of integrity mechanisms, such as message authentication codes, one-way hash functions, and digital signatures, on an IMD can lead the device to accept malicious inputs which can be employed to run a code injection attack. In addition, the lack of integrity mechanisms can facilitate that the alteration of the data stored on the memory of an IMD might not be detected at all or be detected in a distant future [28].

8.5.2   Ensuring availability

Medical IoT device's resources (e.g., data, services, memory, and battery) should be available to legitimate entities at all times. Availability is a crucial security objective for medical IoT devices that should operate always properly as they are dedicated to treat medical conditions of their holders [28]. However, medical IoT devices can be rendered inaccessible either due to active jamming attacks causing blockage of the radio channel or due to DoS attacks (e.g., flooding attacks over the radio channel) causing overload of the target device that can lead to battery/memory depletion. This latter type of DoS attacks is known as resource depletion attacks and their focus is on wasting the device's resources. Although these attacks are very easy to implement (e.g., by sending dummy requests), their impact can be very harmful because if the battery runs out of power, the device will become permanently inaccessible and the holder's health can be at risk. Despite the importance of ensuring the availability of medical IoT devices, the prevention of active jamming attacks and resource depletion attacks is not easy. Specifically, the difficulty in prevention of active jamming attacks lies in the use of the wireless communication channel by the medical IoT devices for communication with the outside world. Moreover, regarding the prevention of resource depletion attacks, it is important to highlight that standard cryptographic solutions do not prevent this type of attacks and relevant solutions applied to sensor networks to address resource depletion attacks are not usually applicable to medical IoT devices due to their more severe resource constraints [28].

Nowadays, the most widely used solution to address resource depletion attacks is the combined use of pattern/behavior analysis and notification systems which inform the holder, through an alarm signal, when particular events occur. Nevertheless, these systems do not prevent the attacks. In fact, they only alert the holder to make him/her aware of unexpected ongoing activity. Then, the holder is responsible to take decisions and react appropriately to the specific notifications. Last but not least, another shortcoming of these systems comes from the fact that these systems base their security on an external device (e.g., holder's smartphones) [28]. Thus, it is clear that more effort should be put into the design and implementation of security solutions to prevent resource depletion attacks against medical IoT devices and ensure their availability.

8.5.3   Ensuring authentication and authorization

Medical IoT devices' data and operations should be accessed only by the requester with the appropriate rights. For example, therapy parameters (e.g., voltage, current, thresholds, and operation mode) can be updated only by the doctors and not by the patient. In this regard, re-programming/update the medical IoT device should be done under the joint supervision of the doctor and a technician (typically from the manu-facturing company of the device) [28]. Authentication and access control mechanisms are used to ensure authenticated and authorized access to information and operations on the medical IoT devices. In particular, these mechanisms target to: (a) verify the validity of the claimed identity (i.e., authentication) of the entity (e.g., patient and doctor) requesting access to data/operation on the medical IoT device, and (b) guarantee that the authenticated entity can only access the data/operation that he/she is authorized for. Thus, requesters will be allowed to view, share, or modify information on medical IoT devices only after his/her successful authentication and authorization. Moreover, these mechanisms will provide restricted access to the device according to the requester's privileges that specify access control information. Consequently, the level of access for each authorized requester can be controlled and thus reducing the risk of intrusion attacks. It is noteworthy to say that access control is also a method to achieve data confidentiality for the stored data on the medical IoT device [28]. A good example of access control mechanisms is the Access Control Lists (ACLs) which constitute an implementation of discretionary access control models based on the access matrix. The ACLs define the operations or data that an authenticated requester is authorized to execute or access, respectively. It is worthwhile to mention that such permissions are permanent once the ACLs are programmed [28].

8.5.4   Ensuring nonrepudiation

Nonrepudiation prevents an entity from denying previous commitments or actions. In particular, when disputes come from the entity denying his/her previous commitments or actions, a means to resolve this situation is essential [20]. In medical IoT devices, the means which is used to resolve these situations is the access log where all operations performed by/on them are stored securely [29]. In other words, auditing helps to mitigate attacks against nonrepudiation. In fact, auditing does not prevent the occurrence of attacks, but it acts as a deterrent means in case that it is appropriately implemented, specifically, if it is not possible for an attacker to compromise the audit log [30]. Therefore, auditing should be complemented with appropriate mechanisms to detect and block attacks against nonrepudiation as well as with mechanisms (e.g., encryption and access control) that will allow the prevention from occurring at first [28].

8.6 Emerging good practice and approaches (standards)

The FDA has issued several sets of guidance demonstrating that medical device cybersecurity is a significant issue. Both the post-market management of cybersecurity in medical devices and the interoperable medical devices contain specific guidance on cybersecurity. The FDA recommends manufacturers use the National Institute of Standards and Technology (NIST) Framework for Improving Critical Infrastructure Cybersecurity which builds on earlier guidance for Industrial Control Systems. Both the pre-market guidance (October 2014) and the post-market guidance (December 2016) recommend manufacturers utilize the NIST Framework for Improving Critical Infrastructure Cybersecurity. This has core functions to guide an organization's cybersecurity activities: Identify, Protect, Detect, Respond, and Recover. The new NIST Cybersecurity Framework (CSF) (draft revision 1.1) places greater emphasis on managing supply chain risk.

The NIST CSF maps reference standards for specific elements and various other frameworks including the Health Insurance Portability and Accountability Act (HIPAA) 1996 (US legislation that provides data privacy and security provisions for safeguarding medical information). The intent is to promote collaboration amongst the medical device and health IT communities and develop a common understanding of cybersecurity vulnerabilities and risk. This will improve assessment of patient safety and public health risks and ability to take timely and appropriate mitigation actions.

EU regulation lags behind the US FDA recommendations. However, the Medical Devices Regulation (MDR) that was published in 2017 significantly enlarged the scope of applicable devices, and defined more stringent post-market surveillance. The draft plans have provisions for vigilance, market surveillance, and reporting in respect of serious incidents and implementing safety corrective actions. In addition, the European-harmonized ISO standard BS EN ISO 62304 "Medical device software — Software life-cycle processes" includes security provisions. The European standard for the application of risk management to medical devices (BS EN ISO 14971:2012) highlights that probabilities are very difficult to estimate for software failure, sabotage, or tampering.

Finally, the international standard used by regulators for medical device surveillance in the post-market phase (DD ISO/TS 19218-1:2011+A1:2013) for sharing and reporting adverse incidents by users or manufacturers does not offer cyber specific categorization. Arguably, a cyber incident could be identified with any of the following categories: computer hardware, computer software, electrical/electronic, external conditions, incompatibility issues, nonmechanical, loss of communications, incorrect device display function, installation, configuration, performance deviation, output issue, protective alarm or fail-safe issue, unintended function—resulting in malfunction, misdiagnosis, or mistreatment, or simply as "other device issue" [17].

8.7 Conclusions and further challenges

Medical IoT devices improve the quality life of patients and, in some cases, play an important role in keeping them alive. The new generation of medical IoT devices is increasingly incorporating more computing and communication capabilities. However, security issues arise as attacks on healthcare have increased exponentially in recent years. By classifying attacks, we can better organize the security countermeasures to be implemented. We have provided a comprehensive overview of the cybersecurity attacks associated to the most recent medical IoT devices and have discussed how, in some cases, the patient's health can be seriously threatened by a malicious adversary. It is therefore evident that security mechanisms have to be incorporated into these devices. Moreover, there is a need for a much deeper study about how to apply security mechanisms for medical IoT devices in the context of Smart Health. Security issues require flexible, context aware, and adaptive security mechanisms. Toward this direction, further cooperation among researchers coming from manufacturing technologies, bioengineering, and computer security are necessary to guarantee both the patient's safety and the privacy and security of the data and communications [28].

Apart from purely engineering solutions, the procedures that both the medical personnel and the patients follow when interacting with the interconnected devices have to be considered, and existing regulations and standards should also be reviewed. However, nowadays these aspects are essentially ignored. Devices must be used responsibly, and users must know various details about its functioning and the possible threats in order to raise security awareness [28]. Concluding, additional measures need to be taken to handle threats and securing the potential information at both the customer and developer ends. Thus, the vision and long-term success of this dynamically growing industry lays in the synergy of researchers, healthcare professionals, and patients [29,30].

References

[1] ITU-T Focus Group on Smart Sustainable Cities. An overview of smart sustainable cities and the role of information and communication technologies. 2014, p. 30. Retrieved from https://www.itu.int/en/ITU-T/focusgroups/ssc/Documents/Approved_Deliverables/TR-Overview-SSC.docx.

[2]   R. Jalali, K. El-khatib, and C. McGregor, "Smart city architecture for community level services through the Internet of Things," in *2015 18th International Conference on Intelligence in Next Generation Networks*, 2015, pp. 108–113.

[3]   M. S. Hossain, G. Muhammad, and A. Alamri, "Smart healthcare monitoring: A voice pathology detection paradigm for smart cities," *Multimed. Syst.*, vol. 25, no. 5, Springer Berlin Heidelberg, 2019, pp. 565–75, doi:10.1007/s00530-017-0561-x.

[4]   M. S. Hossain, "Patient status monitoring for smart home healthcare," in *2016 IEEE International Conference on Multimedia & Expo Workshops (ICMEW)*, 2016, pp. 1–6.

[5]   "Robotic Cardiac Surgery | Johns Hopkins Medicine Health Library." [Online]. Available: https://www.hopkinsmedicine.org/healthlibrary/test_procedures/cardiovascular/robotic_cardiac_surgery_135,11. [Accessed: 30-Jan-2019].

[6]   A. Schneider, "Development of Smart Injection Devices: Insights from the Ypsomate Smart Case Study," *ONdrugDeliveryMagazine*, no. 64, pp. 6–9, February 2016.

[7]   "YpsoDose—Pre-assembled patch injector—Ypsomed Delivery Systems." [Online]. Available: https://yds.ypsomed.com/en/injection-systems/ypsodose.html. [Accessed: 6-Mar-2019].

[8]   "Remote Medical Center—Telemedicine, EHR system—Comarch Healthcare." [Online]. Available: https://www.comarch.com/healthcare/products/remote-medical-care/remote-medical-center/. [Accessed: 06-Mar-2019].

[9]   P. Dario, B. Hannaford, and A. Menciassi, "Smart surgical tools and augmenting devices," *IEEE Trans. Robot. Autom.*, vol. 19, no. 5, pp. 782–792, 2003.

[10]   B. Farahani, F. Firouzi, V. Chang, M. Badaroglu, N. Constant, and K. Mankodiya, "Towards fog-driven IoT eHealth: Promises and challenges of IoT in medicine and healthcare," *Futur. Gener. Comput. Syst.*, vol. 78, pp. 659–676, 2018.

[11]   "WHO | Medical Device—Full Definition," *WHO*, 2018.

[12]   Center for Devices and Radiological Health. "Postmarket Management of Cybersecurity in Medical Devices – Guidance." U.S. Food and Drug Administration, FDA, Dec. 2016, https://www.fda.gov/regulatory-information/search-fda-guidance-documents/postmarket-management-cybersecurity-medical-devices.

[13]   "EU European Medical Device Classification." [Online]. Available: https://www.emergobyul.com/services/europe/european-medical-device-classification. [Accessed: 15-Jan-2019].

[14]   M. Zhang, A. Raghunathan, and N. K. Jha, "Trustworthiness of medical devices and body area networks," *Proc. IEEE*, vol. 102, no. 8, pp. 1174–1188, 2014.

[15]   "Guidelines for Classification of Medical Devices – CE Marking (CE Mark) for Medical Devices – EU Council Directive 93/42/EEC." [Online]. Available: http://www.ce-marking.org/Guidelines-for-Classification-of-Medical-Devices.html. [Accessed: 15-Jan-2019].

[16]   S. C. Risks, "Securing Connected Hospitals."

[17] R. Piggin, "Cybersecurity of medical devices: Addressing patient safety and the security of patient health information," *BSI Group*, 2017, pp. 1–29, doi:10.1002/jclp.22023.

[18] "Confidentiality, privacy and security of health information: Balancing interests – Health Informatics Online Masters | Nursing & Medical Degrees." [Online]. Available: https://healthinformatics.uic.edu/blog/confidentiality-privacy-and-security-of-health-information-balancing-interests/. [Accessed: 21-Nov-2018].

[19] M. M. Hossain, M. Fotouhi, and R. Hasan, "Towards an analysis of security issues, challenges, and open problems in the Internet of Things," in *2015 IEEE World Congress on Services*, 2015, pp. 21–28.

[20] A. J. Menezes, P. C. Van Oorschot, and S. A. Vanstone, "Handbook of Applied Cryptography." Handbook of Applied Cryptography, 1996, doi:10.5860/choice.34-4512.

[21] M. S. Jalali and J. P. Kaiser, "Cybersecurity in hospitals: A systematic, organizational perspective," *J. Med. Internet Res.*, vol. 20, no. 5, p. e10059, 2018.

[22] E. Gaukstern, and S. Krishnan, "Cybersecurity Threats Targeting Networked Critical Medical Devices," *ASEE IL-IN Sect. Conf.*, Aug. 2018.

[23] S. D. Smalley, "Laying a Secure Foundation for Mobile Devices." NDSS (2013).

[24] "Passive attack – Wikipedia." [Online]. Available: https://en.wikipedia.org/wiki/Passive_attack. [Accessed: 24-Nov-2018].

[25] Dec, Oct-, et al. Impersonation and Spoofing Fraud Q4 2015. 2016, pp. 2015–17, http://www.mycert.org.my/en/services/.

[26] "Top 10 Most Common Types of Cyber Attacks." [Online]. Available: https://blog.netwrix.com/2018/05/15/top-10-most-common-types-of-cyber-attacks/. [Accessed: 28-Dec-2018].

[27] P. A. Williams and A. J. Woodward, "Cybersecurity vulnerabilities in medical devices: A complex environment and multifaceted problem." *Med. Devices (Auckl).*, vol. 8, pp. 305–316, 2015.

[28] C. Camara, P. Peris-Lopez, and J. E. Tapiador, "Security and privacy issues in implantable medical devices: A comprehensive survey." Journal of Biomedical Informatics, vol. 55, Academic Press Inc., 1 June 2015, pp. 272–89, doi:10.1016/j.jbi.2015.04.007.

[29] F. Campioni, S. Choudhury and F. Al-Turjman, "Scheduling RFID Networks in the IoT and Smart Health Era." *Journal of Ambient Intelligence and Humanized Computing*, vol. 10, no. 10, Springer Verlag, Oct. 2019, pp. 4043–57, doi:10.1007/s12652-019-01221-5.

[30] F. Al-Turjman and S. Alturjman, "Context-sensitive access in industrial Internet of Things (IIoT) healthcare applications," *IEEE Transactions on Industrial Informatics*, vol. 14, no. 6, pp. 2736–2744, 2018.

Chapter 9

HaLow: registering thousands of low-power sensors in smart cities

Rashid Ali,[1] Nurullah Shahin,[2] Fadi Al-Turjman,[3] Byung-Seo Kim,[4] and Sung Won Kim[5]

IEEE 802.11ah working group (WG) introduced Wi-Fi HaLow (or simply HaLow which is a marketing name of Wi-Fi for low-power devices) as a revision of the long-range Wi-Fi technology for the Internet of Things (IoT) applications in smart cities. Such applications often involve thousands of wireless devices (typically sensors and actuators) connected to a shared wireless channel. Channel access for these thousands of IoT devices significantly affects the performance of the network. Several mechanisms have been proposed to register thousands of low-power sensors and actuators by handling the contention between them. Two of which known as the centralized authentication control (CAC) and the distributed authentication control (DAC), aimed to address the contention reduction during the link set-up process in HaLow. In HaLow, a link set-up process requires much more interests from the researchers because the access point (AP) knows nothing about the connected devices and the mean of control at these stations is very limited. DAC is a self-adaptive device authentication mechanism, whereas CAC requires an algorithm to dynamically control critical parameters, such as transmission slots and channel access period. However, the existing IoT devices registration mechanism in HaLow is based on carrier-sense multiple access with collision avoidance (CSMA/CA), which is not very efficient for the registration of large-scale IoT devices due to its limited binary exponential contention mechanism. In this chapter, we explain both of the device authentication mechanisms, i.e. CAC and DAC, in detail. Later, we discuss one of the authentication mechanism known as hybrid slotted-CSMA/CA-time-division multiple access (TDMA) (HSCT) as our case study that is proposed to overcome the aforementioned issues in current authentication techniques. The HSCT mechanism allows IoT systems in smart cities to register thousands of low-power IoT devices

[1]School of Intelligent Mechatronics Engineering, Sejong University, Seoul, Korea
[2]IT Department, The Central Bank of Bangladesh, Dhaka, Bangladesh
[3]Computer Engineering Department, Antalya Bilim University, Antalya, Turkey
[4]Department of Computer and Information Communication Engineering, Hongik University, Sejong-si, Korea
[5]Department of Information and Communication Engineering, Yeungnam University, Gyeongsan-si, Korea

(sensors and actuators). This chapter also comes up with the analyses of the access period in a single HSCT time slot.

9.1 Introduction

The IoT is aiming to connect numerous low-power sensors and actuators as part of the future network infrastructures [1]. However, there exist many challenges and issues to the current wireless networks incorporating these sensors and actuators [2], for example, energy efficiency, network reliability, and latency in the network in case of a large number of IoT devices connected to a single wireless channel (which requires contention for accessing the channel resources) [3–5]. Designing an appropriate wireless technology considering contention for registration as well as for transmission of data is still an open issue [6–8]. One of the good consideration for such issues is the emergence of Wi-Fi HaLow technology by IEEE 802.11ah WG [9]. HaLow has evolved as an attempt to meet the requirement of registering thousands of low-power IoT devices to the network while sticking to the basic ideology of Wi-Fi, which is also known as wireless local area networks (WLANs). There are many new mechanisms introduced by this WG to the WLAN that address the IoT system peculiarities, as shown in Figure 9.1. One of the mechanism is restricted access window (RAW) [10], which is introduced to register up to 8,000 of IoT devices by grouping and assign them dedicated time slots for registration and transmission. This procedure decreases the contention for the channel by limiting the number of contenders and thus improves transmission reliability. Another mechanism is the

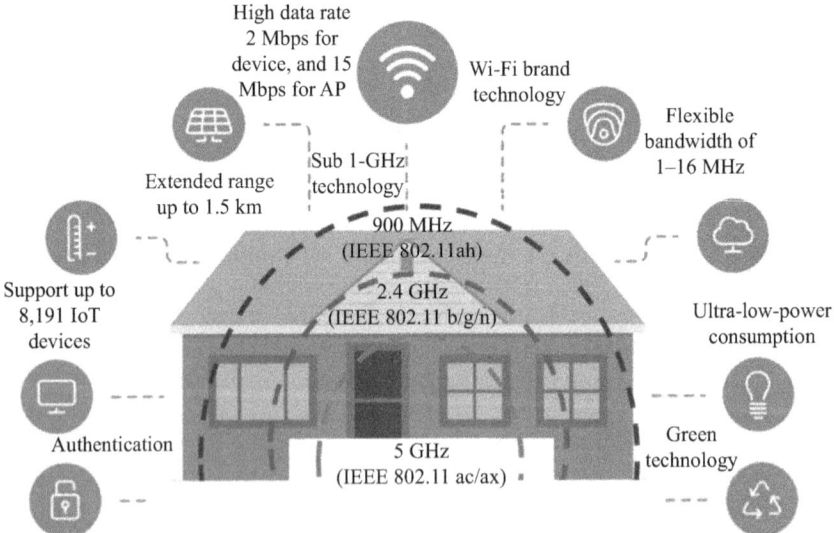

Figure 9.1   HaLow enhancing the capabilities of Wi-Fi in IoT systems

traffic indication map (TIM) segmentation [11], which is used to improve the energy efficiency of a WLAN by grouping the devices and enabling them to receive only group-specific packets.

RAW and TIM in HaLow together will help to make it suitable for massive IoT-based communications. However, in HaLow, a central AP needs to set up a link between AP and node (device) before it utilizes any of the mechanism mentioned above. In the link set-up procedure, an AP learns about the existence of the node, and its capabilities and admissibility to the network. AP informs the node of the network parameters and assigns it a network identifier, known as association identifier (AID) that is used throughout the entire communication procedure with this node. AP uses CSMA/CA mechanism [11] to transmit the management frames for link set-up. CSMA/CA mechanism is a channel access mechanism based on a binary exponential backoff (BEB) [12] algorithm. However, BEB itself introduces numerous challenges of the increase in a collision due to blind contention mechanism [13–15], and thus is only sufficient to provide fast link set-up in smaller networks; whereas, for a high number of contending devices, CSMA/CA suffers from severe collisions and provides poor network performance. This also results in the form of longer link set-up disconnection, which effects on the performance of already connected devices. The importance of link set-up in IoT systems can be seen when a huge number of IoT devices have to re-register if AP serving this group of devices reboot due to any malfunctioning reason [16,17].

IEEE 802.11ah WG (HaLow) proposed two link set-up protocols to reduce the contention among the devices and makes it faster to register [10]. These two are known as CAC and DAC protocols. CAC protocol allows the AP to set a portion of devices periodically for sending a request for link set-up, whereas the DAC protocol allows devices to use random time intervals for their link set-up requests to the AP. Both, CAC and DAC, have a list of parameters to perform their registration procedure. There has been a lot of research work related to the fast link set-up in HaLow networks, such as [18–20], and they have improved the registration of thousands of devices for IoT systems in HaLow. This chapter gives a brief description of CAC and DAC protocols. Later in this chapter, one of the recently proposed massive IoT-devices registration mechanisms, known as HSCT [10], is discussed. We analyse the registration of the possible device in a single slot of the HSCT.

The rest of the chapter is organized as follows. In Section 9.2, we present an overview of IEEE 802.11ah HaLow standard. Section 9.3 provides a brief background on link set-up in HaLow networks including CAC and DAC protocols. In Section 9.4, we describe HSCT mechanism as our case study for registering thousands of low-power IoT devices. Finally, Section 9.5 gives our concluding remarks of the chapter.

9.2 IEEE 802.1ah HaLow: an overview

The purpose and scope of the HaLow (IEEE 802.11ah project) was described in the primary work of a Project Authorization Request (PAR) document [21] work taken by

WGah in 2010. Initially, an amendment to develop a selection procedure was proposed, which would be executed and followed by the WGah until the final draft [22]. For this purpose, WGah adopts channel usage models [23], channel models [24], and channel functional requirements [25]. However, WGah faced numerous technical challenges, such as low-power devices support, long-range transmission, huge number of device registration, and network performance enhancement, most of which have already been resolved. However, there are still some remaining challenges, which are needed to be carefully addressed. Since the coverage of a HaLow AP can be substantially large, the performance of a certain HaLow network can be severely affected by the interference generated by numerous devices as well as neighbouring networks. For the aforementioned issues, an effective solution adopted by WGah is to employ CAC and DAC mechanisms to divide the whole network into several regions and use a device registration approach to spread the transmissions of different devices; whereas standardization on the large number of devices registration is still ongoing.

9.2.1 Use cases for HaLow

There is no doubt that characteristics of HaLow have attracted the researchers in a very short period. Generally, it is described by the WGah in their initial drafts that HaLow use cases include low-power wireless sensor networks and extended range backhaul networks for sensors [23]. These two major use cases help to understand the advantages of HaLow technology for various scenarios.

9.2.1.1 Smart-grid: low-power wireless sensor networks

To be the part of green technology [26], electrical grids [27] start deploying a huge number of sensor devices and meters inside their existing infrastructures, named as smart grids. The functions and characteristics of smart grids are to monitor the real-time status of various electricity consummations and inform the company and customers of their usage status. Typically, the number of devices involved in smart grid is much higher than that in traditional WLANs, and the required transmission range of the involved devices is much wider than that in traditional WLANs [28]. A very simple smart grid scenario is shown in Figure 9.2. The figure shows that a

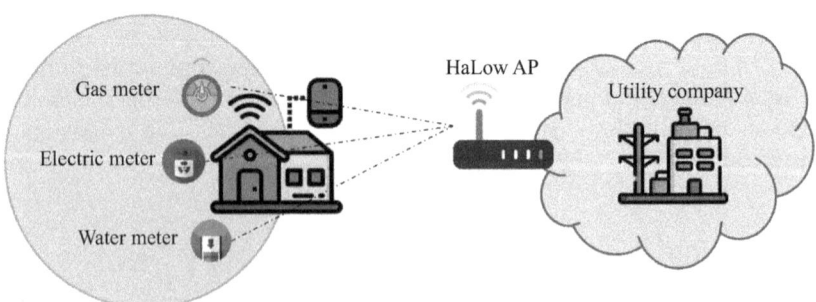

Figure 9.2   HaLow smart-grid use case

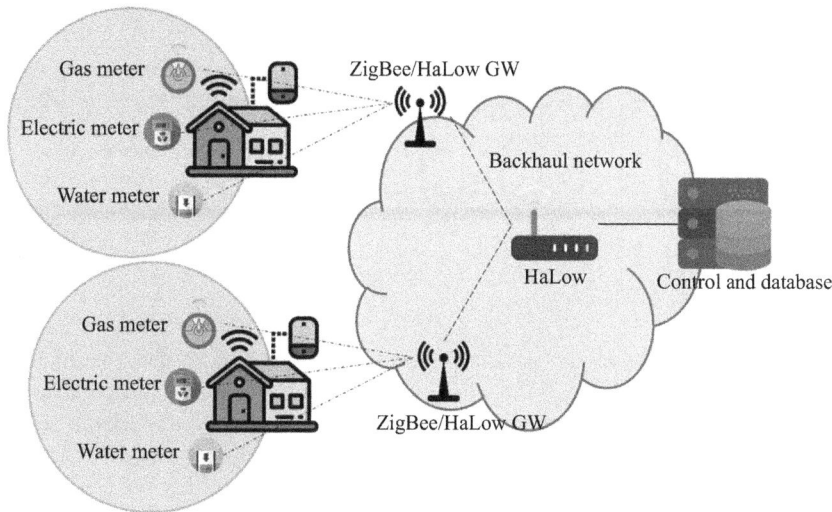

Figure 9.3 HaLow long-range backhaul use case

HaLow AP is placed at the outdoor, and the devices (sensors/meters) are deployed in the surrounding indoor area. For outdoor scenarios, HaLow can support up to 1 km of coverage with a minimum of 100 kbps data rate [23].

9.2.1.2 Long-range backhaul

Another use case is the long-range extended backhaul connection between sensors and AP. In this use case, ZigBee [29] provides communication services to sensors and devices, and HaLow provides a wireless backhaul connection to APs for aggregated traffic transmission. Besides, the extended coverage of HaLow allows a simple network design to link central APs together as shown in Figure 9.3. This figure clearly demonstrates a long-range backhaul network, composed of HaLow AP and gateways, which aggregate and forward the network traffic from connected devices, to remote control and database [30].

9.2.2 HaLow MAC layer

In HaLow medium access control (MAC) layer, some features are enhanced as compared with the WLAN MAC layer, which includes improvements related with support of large number of energy-constrained devices, link set-up mechanisms, and throughput enhancements. Enhanced MAC layer of HaLow technology supports for a large number of low-power associated devices [31].

As described earlier, in HaLow, an AP allocates an AID to each device during the association period, where, an AID is a unique number which indicates its associated devices. The possible number of associated devices of an AP is up to 2,007 in legacy 802.11 standard due to the limited length of the partial virtual bitmap of TIM, where each bit indicates the corresponding device's AID [12]. As described earlier, in HaLow

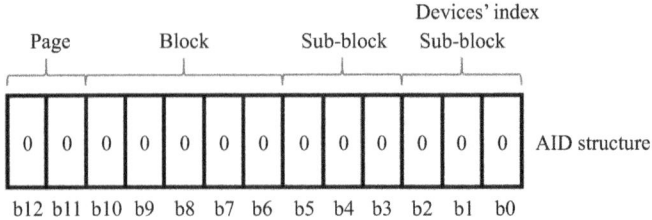

Figure 9.4   HaLow AID structure

network, an AP is likely to be associated with much more devices than that in legacy 802.11 networks, and hence, HaLow has increased the number of supportable devices to meet such expected requirements. For increasing the number of supportable devices, a hierarchical AID structure is defined in HaLow as shown in Figure 9.4. The MAC layer header consists of 13 bits and, thus, the number of devices that can be supported is up to 8,191 ($2^{13} - 1$). It is composed of four hierarchical levels: page, block, sub-block, and device's index in sub-block. That is, each device belongs to a certain sub-block, and each sub-block belongs to a certain block. Similarly, multiple blocks form a page, which is the highest level that can contain up to 2,048 devices. This hierarchical AID structure enables us to indicate more devices' AID with a given length of partial virtual bitmap. For example, when we need to indicate multiple devices, we can simply include them in a block or a sub-block and use the block ID or the sub-block ID to indicate them instead of including all of their AIDs. Furthermore, there could be devices with different traffic types. If we can easily group these devices based on some specific properties, the wireless resource could be utilized more efficiently. There have been many existing research efforts on clustering algorithms based on traffic types. As a well-known principle, hierarchical structure can be adopted to facilitate the grouping so that the characteristic of hierarchical AID structure makes grouping of the devices much easier.

9.3   Link set-up in HaLow networks for massive IoT devices registration

HaLow uses link set-up mechanism very similar to the ordinary IEEE 802.11 WLANs (i.e. Wi-Fi). Therefore, to better understand the link set-up mechanism in HaLow, we briefly overview the basic association procedure in WLANs. In this section, we briefly describe link set-up in infrastructure WLANs (that is a network with centralized AP), focusing on authentication and association procedures as well as on the default channel access procedure.

9.3.1   Link set-up in WLANs

In WLANs, AP and the nodes have to perform two handshakes for the link set-up, authentication, and association as shown in Figure 9.5. Authentication of the devices

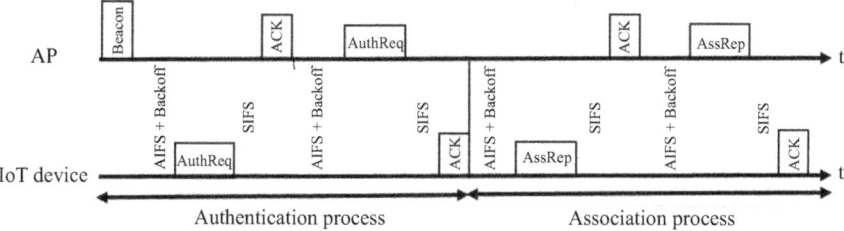

Figure 9.5   Link set-up procedure in WLAN/HaLow networks

starts after a device receives a periodically sent beacon message from the AP, known as Probe Response (ProbResp) frame. The ProbResp message helps devices to learn of the AP and the surrounding network. During the authentication handshake phase, the AP and the devices confirm their identity with the help of security keys exchange. In this procedure, the device sends an Authentication Request (AuthReq) frame to the AP, and in return, AP responds to the device with an Authentication Response (AuthRep) frame. This authentication phase is mandatory even if the network uses an open access system authentication which is unsecured network configuration [32].

In the second handshake phase, which is the association phase, the device informs the AP about its capabilities, such as supported modulation and coding schemes, and channels to use. Then the AP sets the channel access parameters for the device by assigning an AID to the device. The AP to identify the device for management frames transmission, such as beacon messages, uses this AID. An Association Request (AReq) frame is sent by the device to associate itself with an AP, to which the AP replies back with an Association Response (ARep) frame. After receiving an ARep frame, the device may transmit and receive further data frames [33].

Enhanced Distributed Channel Access (EDCA), which is an enhanced form of CSMA/CA, uses AuthReq, AuthRep, AReq, and ARep frames for the authentication and association process. Every device has to sense the channel before the transmission of any above-mentioned frame. A random backoff mechanism is used to differ the channel for collision avoidance. After observing the channel as idle for an arbitrary interframe space (AIFS), the device draws a random value from a contention window (CW) [34]. The device keeps sensing the channel for the selected number of backoff slots, and the channel is accessed only if the backoff value reaches zero successfully. Every time, the transmission is detected at the channel, the device freezes the backoff and waits for AIFS to resume. The receiving device sends back an acknowledgment (ACK) to the transmitter after a short interframe space (SIFS). The SIFS period is very short so that other devices may not consider it as an idle channel for the new transmission. The only way to determine the collision in a wireless network is to know if ACK is received or not. If ACK is not received within the AckTimeout period, the transmitted data are considered as the collision and the channel is accessed again for the transmission. This time, device exponential doubles the size of previous CW, known as BEB mechanism [13,14]. The unsuccessful transmission attempts are carried on until the retry limit is also reached. Once a retry limit is reached, the packet is dropped and no longer is transmission attempt done. The standard channel access procedure of CSMA/CA is shown in Figure 9.5.

However, in an IoT network, thousands of IoT devices are connected to a single AP. AP starts the authentication procedure for all of them by triggering a beacon or Probe message. ProbResp from the devices results in a high contention between numerous AuthReqs by the devices. Since, during the authentication phase, the capabilities of the IoT devices are not known, there are many control mechanisms introduced in IEEE 802.11ah, such as RAW. However, RAW is unsuitable and the devices have to rely on the basic random channel access mechanism (i.e. CSMA/CA) to transmit its AuthReq messages; whereas the performance of current CSMA/CA significantly degrades with increase in the number of transmitting devices [35]. For that reason, as we discussed earlier, the HaLow introduces CAC and DAC authentication control protocols which allow the AP to limit contention for channel access during the authentication phase.

9.3.2   *Distributed authentication control*

In the DAC protocol, beacon intervals (BIs) are the intervals between consecutive beacon messages. BI is further divided into sub-intervals, known as Authentication Control Slots (ACSs) [36]. The devices use a large-scale truncated BEB procedure to reduce the contention. There are m number of BIs and l number of ACSs used for authentication attempts. Thus, an IoT device randomly chooses a BI and an ACS within its chosen BI. At the start of the ACS period, the device attempts to transmit an AuthReq. For this purpose, the device uses CSMA/CA to access the channel. If the AuthReq fails or it does not receive an authentication response message within the specified timeout AuthenticateFailureTimeout, it increments r number of authentication attempts and regenerates m and l. For each authentication attempt a, m is drawn from interval $[0, I_a]$ and l is drawn from $[0, L]$, where L is determined as $L = \frac{BI}{T_{ACS}}$, and I_a is determined as:

$$I_a = \begin{cases} I_{min}, & a = 0 \\ min\{2 \times I_{(a-1)}, I_{max}\}, & a > 0 \end{cases}. \tag{9.1}$$

The ACS period T_{ACS} and both of the interval limits I_{min} and I_{max} are determined by the AP and is advertised with the help of beacon messages, as shown in Figure 9.6. The maximum possible value for T_{ACS} is 127 ms due to the standard frame format [18], whereas maximum possible value for I_{max} is 255 intervals. Selecting optimal values for these parameters is still an open issue for the DAC mechanism. A blind increase in the size of intervals (i.e. I_a) may increase the network latency while restricting it to a smaller size can increase the probability of collision in the network.

9.3.3   *Centralized authentication control*

In contrast to the DAC protocol, CAC allows an AP to dynamically change the portion of devices that are allowed to send AuthReq in the current BI. In this process, beacons and probe messages of AP carry the authentication threshold parameters, an integer

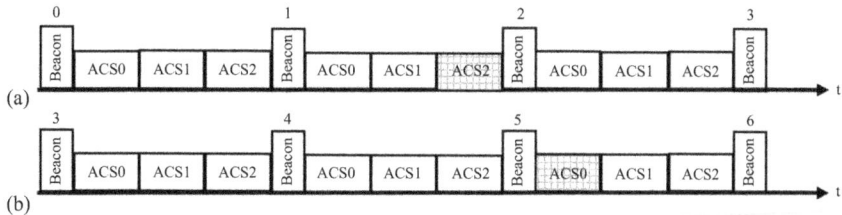

Figure 9.6   *DAC authentication procedure for IoT devices. (a) In the first attempt, a device awakes at beacon 0, and randomly draws m = 1 and l = 2, which results in a failure. (b) In second attempt, a device randomly draws m = 3 and l = 0 and counts beacons starting from 2*

value ranging 0 from 1,023, selected by the AP. The device randomly picks a value between this range and waits for a frame with an authentication threshold. If this value is greater than or equal to the threshold, the device waits for the next. If the value is less than the threshold, the device makes ready to send its AuthReq and content for the channel. If AuthRep is not received within AuthenticateFailureTimeout period, the device waits for a new frame with another appropriate threshold for new AuthReq. In this procedure, an AP controls the number of devices trying to set-up the link, by changing the threshold adaptively in the beacons/probe messages. While on the other hand, IoT devices generate random values only once after being switched on and may generate it again only after a successful AuthRep frame. Hence, the randomly generated number is not updated during a link set-up procedure [10]. In this procedure, the standard does not specify the way for the AP to decide the authentication threshold value. However, this threshold value has a significant impact on the link set-up in terms of latency. Specifically, to minimize the link set-up latency in IoT systems, the AP can set authentication threshold to its maximum size when the number of IoT devices is very small; but, when the number of IoT devices is very large, the AP should increase the threshold gradually from its minimum to its maximum size, which makes a small portion of devices to access the channel at once.

9.4   Case study: hybrid slotted-CSMA/CA-TDMA

As we have seen that one of the issues faced by the registration process with the increase in the number of IoT devices in HaLow is its contention-based CSMA/CA mechanism. Due to the proportional increase in low-power IoT devices for contention, there is a limitation in performance enhancement for CSMA-based IoT systems. In this section, we discuss one of the hybrid approaches, known as HSCT [10], which utilizes the benefits of both contention-based CSMA and contention-free (reservation-based) TDMA mechanisms for efficient exchange of link set-up frames in HaLow networks. Their proposed mechanism is also efficient for the optimal parameters selection of CAC protocol.

In HSCT, an IoT network of N low-power connected devices to an AP is considered. In the considered topology, each IoT device can send/receive management frames (authentication and association) from the AP. A unique ID is assigned to each IoT device. The HSCT assumes that each device is always willing to proceed for authentication and association from the AP. The transmission time is divided into constant BI slots (S_{BI}) which are further divided into three parts: a beacon period (T_B), slotted-CSMA/CA period (T_{SCP}), and the third one slotted-TDMA period (T_{STP}), as shown in Figure 9.7. As shown in the figure, in the beginning, the AP sends broadcast beacon frames to all IoT devices in the range as part of the T_B. These beacons are

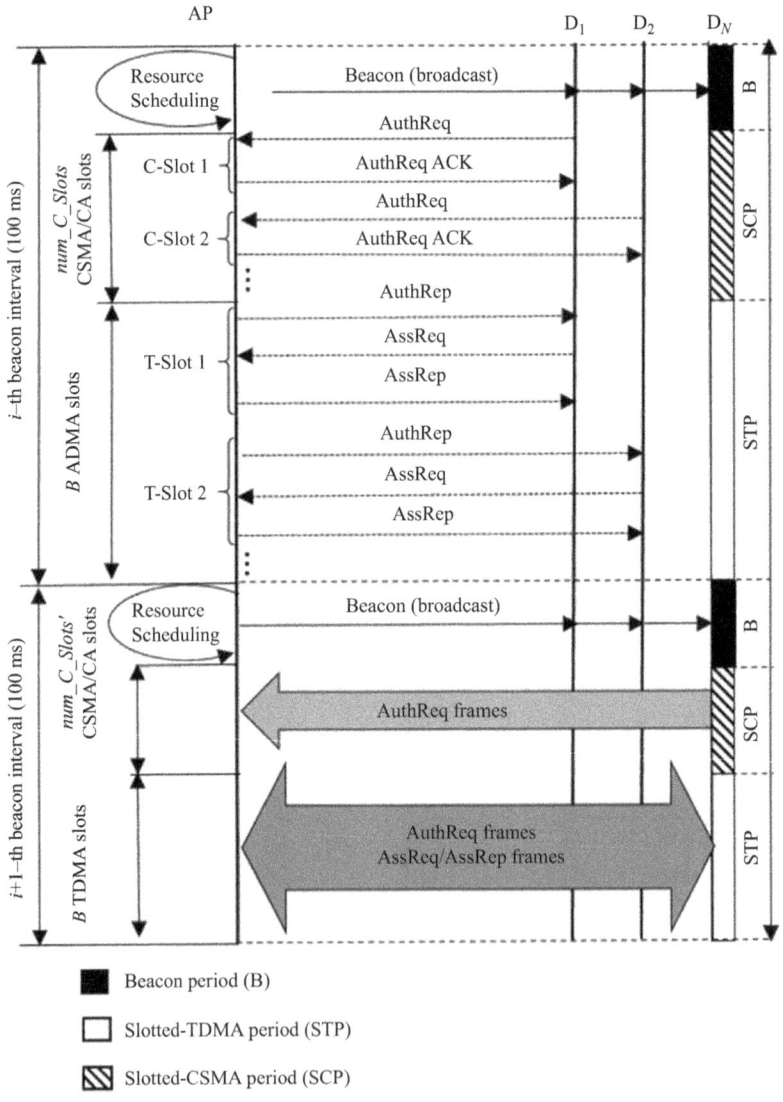

■ Beacon period (B)

☐ Slotted-TDMA period (STP)

▨ Slotted-CSMA period (SCP)

Figure 9.7   IoT device registration in HSCT [10]

used to inform the devices about the beginning of T_{SCP}, number of total slots (n_{SCP}) in the T_{SCP}, the duration of each slot, and the number of total time slots in a beacon (T_B). Along with these time duration information, devices are also informed about the threshold value for authentication control, IDs of unregistered devices, and AIDs of registered devices.

In HSCT, devices are allowed to transmit their AuthReq frames within the T_{SCP} of each T_B using a dynamic slotted CSMA/CA mechanism [37]. Dynamic slotted-CSMA/CA mechanism differs from traditional CSMA/CA because it divides contention period into multiple mini-slots, known as C-slots, where only a group of specific IoT devices are allowed to access the channel. In this mechanism, AuthReq frames are permitted only during a preoccupied T_{SCP}, which means other devices (which are not allowed to transmit in current T_B) have to wait for the next T_{SCP} in the next T_B. In dynamic slotted-CSMA/CA mechanism, SIFT distribution [37] is used to pick randomly generated backoff value from a fixed range of CW. The AuthReq frames are only transmitted once backoff value reaches zero. However, every device have to check before it transmits AuthReq frame that the remaining time in a T_{SCP} is much enough to transmit an AuthReq frame successfully, including AuthReq acknowledgment (AuthReqAck), and the related AIFS, SIFS, and guard time (T_{GI}).

Therefore, from the above definition, we know that during the registration phase all the devices around the AP are the part of either an access group, which comprises the devices that are allowed to send an AuthReq in current T_{SCP}, or the differed group, which comprises the devices that must wait for the next T_{SCP}. The first group of devices randomly select one C-slot from available CSMA/CA slots defined in a single beacon period. It is anticipated that higher numbers of successful AuthReq are possible if contention is at an optimum level with an optimal number of AuthReqs, and the T_{SCP} is long enough to handle the optimal number of AuthReqs. By providing more time for the T_{SCP} in the fixed T_B, the duration of the T_{STP} will be decreased, and the transmission time for successful devices will be reduced. Thus, there is a trade-off between the duration of the T_{SCP} and the T_{STP} in a fixed T_B.

9.4.1 SIFT geometric distribution

One of the different ways adopted by HSCT in its CSMA/CA is the use of SIFT geometric distribution instead of uniform random distribution. Therefore, in this section, we briefly define the SIFT geometric distribution proposed in [37]. The uniform random distribution in traditional CSMA/CA provides the same probability that a device may collide in the network. Even though, in this distribution all the devices possess the same probability of collision, a network may experience high contention at the beginning of each C-Slot period due to equivalent probable values. SIFT is a fixed window CSMA/CA protocol, in which a distribution function $f(p_r)$ is defined as the probability of selecting rth backoff slot, and is given by:

$$f(p_r) = \frac{((1-\alpha) \times \alpha^K)}{(1-\alpha^K)} \times \alpha^r, \quad r = 1, 2, 3, K \tag{9.2}$$

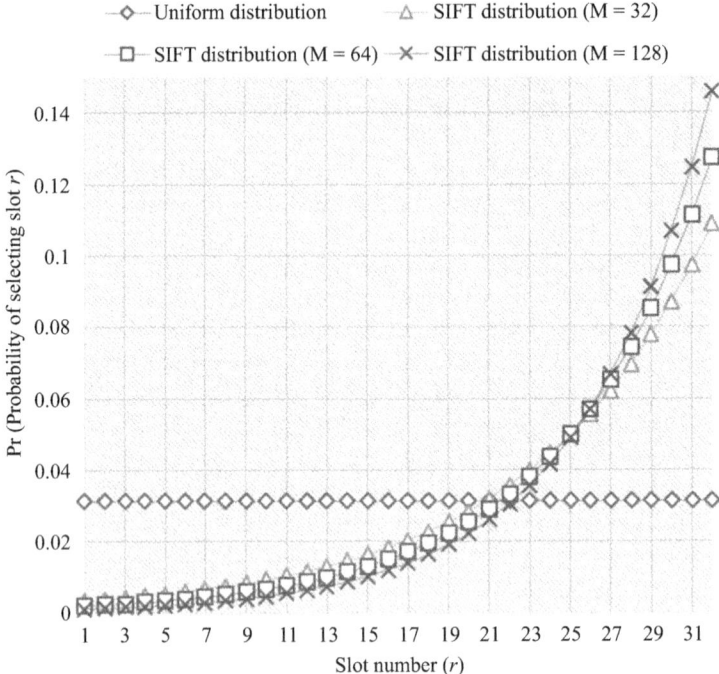

Figure 9.8   Comparison of probability of slots selection in uniform distribution and SIFT geometric distribution [10]

where $\alpha = M^{\frac{-1}{(K-1)}}$ is a distribution parameter ($0 < \alpha < 1$) for M maximum number of contenders. In Figure 9.8, it is shown that using SIFT distribution allows to distribute the IoT devices at the end of the group instead of uniformly distributed over the group. The core benefit of this distribution scheme is to allow as less as possible contenders fall in the initial slots, while most of the devices are placed latterly in the slots, as shown in Figure 9.8. This reduces the probability of collision in the initial backoff slots, thus making most of the transmission successful.

Hybrid slotted-CSMA/CA-TDMA

One of the issues faced by the registration process with the increase in the number of IoT devices in HaLow is its contention-based CSMA/CA mechanism. Due to the proportional increase in low-power IoT devices for contention, there is a limitation in performance enhancement for CSMA-based IoT systems. In this section, we discuss one of the hybrid approach, known as HSCT [10], which utilizes the benefits of both contention-based CSMA and contention-free (reservation-based) TDMA mechanisms for efficient exchange of link set-up frames in HaLow networks. There proposed mechanism is also efficient for the optimal parameters selection of CAC protocol.

9.4.2 Expected number of AuthReq frames in a C-Slot

It is of important interest to know how many AuthReq frames can possibly be transmitted in a single C-Slot. Authors in [10] formulated a closed-form Markov chain analytical model for the expected number of AuthReq frames ($E[AuthReq^i]$) in the ith C-Slot. According to the authors, if $E[B]$ is the expected number of backoff slots and $E[C]$ is the transmission trails faced by one AuthReq frame for the successful transmission, the transmission probability τ for a tagged device can be expressed as follows:

$$\tau = \frac{E[C]}{E[B] + E[C]} \tag{9.3}$$

This equation is an approximation based on renewal reward theory [38], and the values of $E[B]$ and $E[C]$ can be determined as follows:

$$E[B] = \sum_{(j=1)}^{m} E[b_j] \times p^{(m-1)} \tag{9.4}$$

$$E[C] = \sum_{(j=1)}^{m} p^{(m-1)} \tag{9.5}$$

where $E[b_j]$ is the expected number of backoff slots in jth backoff stage for m maximum backoff stages with channel collision probability p. As we know that the HSCT follows a geometric distribution, $E[b_j]$ thus can be determined as follows:

$$E[b_j] = \sum_{(r=1)}^{K} r \times p_r \times (1 - p_r)^{r-1} \tag{9.6}$$

where p_r is probability of selecting slot r according to the SIFT as defined in previous section. Thus Equation (9.3) can be re-written as:

$$\tau = \frac{\sum_{(j=1)}^{m} p^{(m-1)}}{\sum_{(j=1)}^{m} E[b_j] \times p^{(m-1)} + \sum_{(j=1)}^{m} p^{(m-1)}} \tag{9.7}$$

The time slot during T_{SCP} can have one of three states:

1. A slot may be idle if there is no transmission by any of the N IoT devices, with slot duration δ, and can be expressed as:

$$a_{SCP} = (1 - \tau)^{N-1} \tag{9.8}$$

2. A slot may contain a frame transaction either successful or collided. The duration of such slot is equal to transmission of one AuthReq frame, and can be written as:

$$T_{oneAuthReq} = T_{AuthReq} + T_{ACK} + SIFS + AIFS \tag{9.9}$$

3. The last time instant until when the AuthReq transmission attempt can be allowed is known as conflict period (T_C). If a frame transaction begins during the interval $[T_C, T_{OneAuthReq}]$, before the next CSMA slot, and we know that $T_{OneAuthReq} \leq T_C$,

next CSMA slot will arrive during the SIFS period following the frame transaction. Thus, the frame transaction slot finishes earlier then $T_{OneAuthReq}$, let $T'_{OneAuthReq}$ is the duration of such slot. If such condition happens, the beginning of the frame transaction is distributed inside $[T_C, T_{OneAuthReq}]$, with the expected duration as:

$$T'_{OneAuthReq} = \frac{T_{OneAuthReq} + T_C}{2} \tag{9.10}$$

Since the frame transaction can only begin before conflicted period, T_C, so the probability $T'_{OneAuthReq}$ can be $\frac{T_{OneAuthReq} - T_C}{T_B - T_C}$, and thus second state b_{SCP} and third state c_{SCP} can be expressed as:

$$b_{SCP} = 1 - a_{SCP} - c_{SCP} \tag{9.11}$$

$$c_{SCP} = (1 - a_{SCP}) \times \frac{T_{OneAuthReq} - T_C}{T_B - T_C} \tag{9.12}$$

Now the expected slot duration within the access period T_{SCP} is expressed as:

$$E[\Theta_{SCP}] = a_{SCP} \times \delta + b_{SCP} \times T_{OneAuthReq} + c_{SCP} \times T'_{OneAuthReq} \tag{9.13}$$

Another possibility for a transmission is to be started inside the time interval $[T_C, T_C + T_{OneAuthReq}]$ before the arrival of the next CSMA-slot. Such transmission will end inside the conflict time, which makes the vulnerable time T_V smaller than conflict time T_C on average. In this case, we approximate that the starting point of the transmission is uniformly distributed inside $[T_C, T_C + T_{OneAuthReq}]$, so the expected vulnerable time can be obtained as $T_V = \frac{T_C}{2}$. However, if there is no transmission within this interval, the $T_V = T_C$. The number of idle slots Θ in $[T_C, T_C + T_{OneAuthReq}]$ can be written as:

$$\Theta_{T_{OneAuthReq}} = \frac{T_{OneAuthReq}}{\delta} \tag{9.14}$$

and the probability of no transmission in this interval is $a_{SCP}^{\Theta_{T_{OneAuthReq}}}$. Thus, the expected length of the T_V can be expressed as:

$$T_V = a_{SCP}^{\Theta_{T_{OneAuthReq}}} \times T_C + \left(1 - a_{SCP}^{\Theta_{T_{OneAuthReq}}}\right) \times \frac{T_C}{2} \tag{9.15}$$

which can be further solved as:

$$T_V = \left(1 + a_{SCP}^{\Theta_{T_{OneAuthReq}}}\right) \times \frac{T_C}{2} \tag{9.16}$$

Now the expected length of access period within a CSMA-slot can be obtained by $T_{SCP} = T_i - T_V$, as shown in Figure 9.8, and the expected number of successful registrations in a single slot can be derived from:

$$E[AuthReq^i] = \frac{E[T_{SCP}]}{E[\Theta_{SCP}]} \tag{9.17}$$

Therefore, the expected number of AuthReqs in a single $E[T_{SCP}]$ can be derived as:

$$E[\Theta_{T_{SCP}}] = E[AuthReq^i] \times num\_C\_Slots \tag{9.18}$$

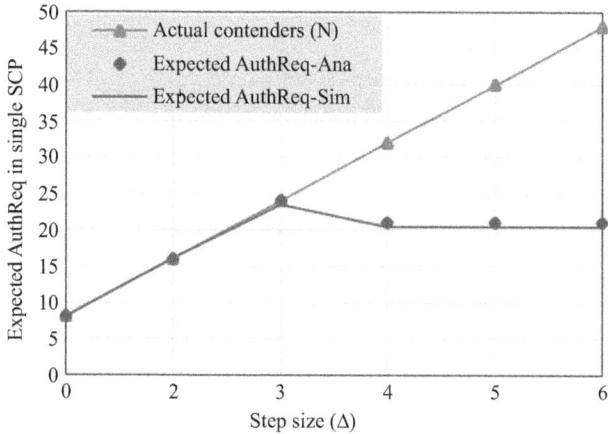

Figure 9.9   Analysis and simulation results for the expected number of AuthReq frames in a single slotted CSMA-CA procedure [10]

where num_C_Slots is the number of C-Slots in a single T_{SCP}. The average number of AuthReqs in one SCP, with a variation of the number of contenders (M) is plotted in Figure 9.9. The analytical results are obtained according to Equations (9.1) to (9.18).

9.5   Conclusions

IEEE 802.11ah WG introduced Wi-Fi HaLow (or simply HaLow which is a marketing name of Wi-Fi for low-power devices) as a revision of the long-range Wi-Fi technology for the IoT applications in smart cities. Such applications often involve thousands of wireless devices (typically sensors and actuators) connected to a shared wireless channel. Channel access for these thousands of IoT devices significantly affects the performance of the network. Several mechanisms have been proposed to register thousands of low-power sensors and actuators by handling the contention between them. In this chapter, we introduce IoT devices' registration in HaLow networks. Two states of the art device registration (association and authentication) mechanisms, known as DAC and CAC, are defined for this purpose. Currently, DAC and CAC seem inefficient in terms of supporting registration of thousands of low-power IoT devices in HaLow networks due to traditional CSMA/CA mechanism. One of the limitations of CSMA/CA for numerous device registration is the use of uniform probability distribution for collision avoidance. One of the solutions to this limitation recently has been published with the name of HSCT which provides efficient and scalable device registration procedure for a large number of low-power IoT devices (i.e. up to 8,192). HSCT proposes the use of multiple CSMA/CA access slots to make the network compatible and avoid congestion in the presence of a large number of IoT devices. This chapter introduces the use of HSCT for massive low-power IoT device registration.

References

[1] Al-Turjman F. The road towards plant phenotyping via WSNs: An overview. Elsevier Computers & Electronics in Agriculture. 2019;161(1):4–13.

[2] Al-Turjman F, and Abujubbeh M. IoT-enabled smart grid via SM: An overview. Elsevier Future Generation Computer Systems. 2019;96(1):579–590.

[3] Al-Turjman F. Mobile couriers' selection for the smart-grid in smart cities' pervasive sensing. Elsevier Future Generation Computer Systems. 2018;82(1):327–341.

[4] Al-Turjman F, and Malekoo A. Smart parking in IoT-enabled cities: A survey. Elsevier Sustainable Cities and Societies. 2019;49(1):1–20.

[5] Hasan MZ, and Al-Turjman F. Analysis of cross-layer design of quality-of-service forward geographic wireless sensor network routing strategies in Green Internet of Things. IEEE Access. 2018;6(1):20371–20389.

[6] Al-Turjman F. 5G-enabled devices and smart-spaces in social-IoT: An overview. Elsevier Future Generation Computer Systems. 2019;92(1): 732–744.

[7] Alchihabi A, Dervis A, Ever E, *et al*. A generic framework for optimizing performance metrics by tuning parameters of clustering protocols in WSNs. Springer Wireless Networks. 2019;25(3):1031–1046.

[8] Deebak D, Ever E, and Al-Turjman F. Analyzing enhanced real-time uplink scheduling algorithm in 3GPP LTE-advanced networks using multimedia systems. Transactions on Emerging Telecommunications. 2018;29(10):e3443.

[9] Khorov E, Lyakhov A, Krotov A, *et al*. A survey on IEEE 802.11 ah: An enabling networking technology for smart cities. Computer Communications. 2015;(58):53–69.

[10] Shahin N, Ali R, and Kim Y. Hybrid slotted-CSMACA-TDMA for efficient massive registration of IoT devices. IEEE Access. 2018;6:18366–18382.

[11] Ali R, Qadri YA, Zikria YB, *et al*. Q-learning-enabled channel access in next-generation dense wireless networks for IoT-based eHealth systems. EURASIP Journal on Wireless Communications and Networking. 2019;178:1–12.

[12] Ali R, Kim SW, Kim BS, *et al*. Design of MAC layer resource allocation schemes for IEEE 802.11ax: Future directions. IETE Technical Review. 2016;35(1):28–52.

[13] Ali R, Shahin N, Bajracharya R, *et al*. A self-scrutinized backoff mechanism for IEEE 802.11ax in 5G unlicensed networks. Sustainability. 2018;10(4):1–15.

[14] Ali R, Shahin N, Kim Y, *et al*. Channel observation-based scaled backoff mechanism for high-efficiency WLANs. Electronics Letters. 2018;54(10): 663–665.

[15] Shahin N, Ali R, Kim SW, *et al*. Cognitive backoff mechanism for IEEE802.11ax high-efficiency WLANs. KICS Journal of Communications and Networks. 2019;21(2):158–167.

[16] Al-Turjman F, and Al-Turjman S. Confidential smart-sensing framework in the IoT era. The Journal of Supercomputing. 2018;76(10):5187–5198.

[17] Al-Turjman F, Ever E, Zikria YB, *et al.* SAHCI: Scheduling approach for heterogeneous content-centric IoT applications. IEEE Access. 2019;7: 80342–80349.

[18] Bankov D, Khorov E, and Lyakhov A. The study of the centralized control method to hasten link set-up in IEEE 802.11 ah networks. 21th European Wireless Conference, Budapest, Hungary. 2015; pp. 1–6.

[19] Bankov D, Khorov E, and Lyakhov A. The study of the distributed control method to hasten link set-up in IEEE 802.11ah networks. The 2016 XV International Symposium Problems of Redundancy in Information and Control Systems (REDUNDANCY), St Petersburg, Russia. 2016; pp. 13–17.

[20] Bankov D, Khorov E, Lyakhov A, *et al.* Fast centralized authentication in Wi-Fi HaLow networks. The 2017 IEEE International Conference on Communications (ICC), Paris, France. 2017; pp. 1–15.

[21] Halasz D. Sub 1 GHz license-exempt PAR and 5C. IEEE 80211-10/0001r13. 2010; Available from: https://mentor.ieee.org/802.11/dcn/10/11-10-0001-13-0wng-900mhz-par-and-5c.doc

[22] Halasz D, and Vegt R. IEEE 802.11ah proposed selection procedure. IEEE 80211-11/0239r2. 2011; Available from: https://mentor.ieee.org/802.11/dcn/11/11-11-0239-02-00ah-proposed-selection-procedure.docx.

[23] Vegt R. Potential compromise for 802.11ah use case document. IEEE 80211-11/0457r0. 2011; Available from: https://mentor.ieee.org/802.11/dcn/11/11-11-0457-00-00ah-potentialcompromise-of-802-11ah-use-case-document.pptx.

[24] Porat R, Yong SK, and Doppler K. TGah channel model – proposed text. IEEE 80211-11/0968r3. 2011; Available from: https://mentor.ieee.org/802.11/dcn/11/11-11-0883-01-00ah-channel-model-text.docx

[25] Cheong M. TGah functional requirements and evaluation methodology. IEEE 80211-11/0905r5. 2012; Available from: https://mentor.ieee.org/802.11/dcn/11/11-11-0905-04-00ah-tgah-functional-requirements-and-evaluation-methodology.doc.

[26] Al-Turjman F, Kamal A, Rehmani MH, *et al.* The Green Internet of Things (G-IoT). Wireless Communications and Mobile Computing. 2019;6059343: 1–2.

[27] Al-Turjman F, Al-Turjman C, Din S, *et al.* Energy monitoring in IoT-based ad hoc networks: An overview. Computers & Electrical Engineering. 2019;76:133–142.

[28] Kizilkaya B, Caglar M, Al-Turjman F, *et al.* Binary search tree based hierarchical placement algorithm for IoT based smart parking applications. Elsevier Internet of Things. 2019;5:71–83.

[29] Std I. IEEE 802.15.4g-2012. Part 15.4: Low-rate wireless personal area networks (LR-WPANs) amendment 3: Physical (PHY) specifications for low-data-rate, wireless, smart metering utility networks. IEEE. 2012.

[30] Al-Turjman F, Ever E, and Zahmatkesh H. Small cells in the forthcoming 5G/IoT: Traffic modelling and deployment overview. IEEE Communications Surveys & Tutorials. 2018;21(1):28–65.

[31]   Al-Turjman F. A novel approach for drones positioning in mission critical applications. Transactions on Emerging Telecommunications Technologies. 2019;e3603:1–13.

[32]   Bankov D, Khorov E, Lyakhov A, *et al.* What is the fastest way to connect stations to a Wi-Fi HaLow network? Sensors. 2018;18(9):2744.

[33]   Alabady SA, Al-Turjman F, and Din S. A novel security model for cooperative virtual networks in the IoT era. International Journal of Parallel Programming. 2018; 1–16.

[34]   Al-Turjman F. Fog-based caching in software-defined information-centric networks. Computers & Electrical Engineering. 2018;69:54–67.

[35]   Ali R, Shahin N, Zikria YB, *et al.* Deep reinforcement learning paradigm for performance optimization of channel observation-based MAC protocols in dense WLANs. IEEE Access. 2019;7(1):3500–3511.

[36]   Demir SM, Al-Turjman F, and Muhtaroglu A. Energy scavenging methods for WBAN applications: A review. IEEE Sensors Journal. 2018;18(16): 6477–6488.

[37]   Tay YC, Jamieson K, and Balakrishnan H. Collision-minimizing CSMA and its applications to wireless sensor networks. IEEE Journal of Selected Areas in Communications. 2004;22(6):1048–1057.

[38]   Zhang R, Cai L, and Pan J. Performance analysis of reservation and contention-based hybrid MAC for wireless networks. IEEE ICC. 2010; pp. 1–5.

Chapter 10

Statistical analysis of low-power sensor motes used in IoT applications

Ali Cevat Tasiran[1] and Burak Kizilkaya[2]

Nowadays, Internet of Things (IoT) applications with low-power sensor motes are becoming more and more popular. Environment monitoring and disaster surveillance applications make use of low-power sensors. Energy is one of the most important metrics in such applications. Low-power sensor motes are used to create more energy-efficient applications. Many new architectures and platforms are proposed to support low-power IoT. Although there are many platforms and approaches, the research in the area to analyze low-power wireless sensor networks (WSNs) in terms of energy consumption is not sufficient. Studies from literature propose methods to estimate lifetime, yet statistical analysis with observed data is missing. To analyze such systems, we apply statistical analysis to the data set from *"Intel Berkeley Research Lab"*. Data set includes 35 days of Mica2Dot sensor data including sensor readings and voltage values. The main objective is to analyze effects of environmental variables such as temperature and humidity on lifetime of a sensor node. To understand the data, descriptive analysis is conducted. Some statistical models like linear regression and ordered logit regression are used and results are discussed in detail.

10.1 Introduction

WSNs, wireless multimedia sensor networks (WMSNs) and IoT paradigm have become complementary to each other in especially environment monitoring systems. Environment monitoring or surveillance systems have gained popularity and become easily applicable with the help of IoT and developing technologies. Development in sensor networks makes easier to create smart home, smart city, or smart environment applications [1]. There are many application areas of new developing technology such as military, surveillance, health care, and other smart applications [2,3].

On the other hand, there are many constraints of such systems like fault tolerance, production and deployment costs, and environmental constraints. One of the main constraints except the ones figured out is energy consumption since the sensors

[1]Economics, Middle East Technical University, Northern Cyprus Campus, Cyprus
[2]Computer Engineering, Middle East Technical University, Northern Cyprus Campus, Cyprus

running on battery [4,5]. Energy consumption of the sensors is a hot topic among researchers where many new mechanisms or approaches are proposed. Energy consumption or minimizing the energy consumption of the sensors or IoT applications is important area in the perspective of sustainable environment. As discussed in [6], Information and Communications Technology (ICT) application can become low-carbon enabler by creating energy-efficient systems or they can become power-drainer by not giving enough attention to energy efficiency while creating new IoT systems. In addition, lifetime and reliability analysis of the WSNs is common and popular among researchers in the area. There are several tools and techniques to analyze sensor nodes. Most of the studies in the literature use simulation or analytical modeling as a tool to analyze the sensor lifetime and reliability [7–11]. However, statistical analysis with observed data is rare. Also, there are some studies which discusses energy efficiency and the factors that affect the efficiency of sensor networks [10,12]. However, most of them use comparison techniques or comparative analysis to show how efficient their system or approach. Rather than comparative or descriptive analysis of the system, applying more valuable statistics such as time series analysis, survival analysis, etc., to such systems can draw more valuable conclusions in terms of lifetime analysis of low-power wireless sensors.

In this chapter, we use regular regression technique to examine the lifetimes of sensors. Having information out of analysis gives the control on variables which affect the lifetime. In addition, it makes easier to propose new approaches to increase lifetime of the entire network. Since we have variables which affect the lifetime of the sensor, it will be easier to minimize the energy consumption and maximize the lifetime of the sensors. In this study, some statistical techniques are used to describe and understand data. The results of conducted analysis are discussed in details in the following sections.

The rest of this chapter is organized as follows. Related works in the literature are discussed in Section 10.2. Section 10.3 presents and explains the data which are used in analysis and gives findings of descriptive analysis, linear regression, and ordered logit regression results. Section 10.4 concludes the study and gives recommendations about future research directions.

10.2　Literature review

Lifetime analysis is a popular area for WSNs in the literature. With wide use of low-power sensors, it became more and more important research area [13–16]. In the literature, analysis are done using several methods. It can be categorized as analytical modeling, experiment, and simulation as summarized in Table 10.1.

In the study of Chen *et al.* [7], general formula for lifetime analysis is derived using analytical modeling. Proposed formula identifies two key variables which affect the lifetime of the network which are channel state and residual energy of the sensor. Using similar approach, Duarte-Melo *et al.* [8] proposed a mathematical formulation to estimate energy consumption and lifetime of a sensor node based on a clustering mechanisms with parameters related to sensing field like size and distance. Given

Table 10.1   Literature summary

Study	Year and Reference	Method used
Chen, Yunxia, and Qing Zhao, "On the lifetime of wireless sensor networks."	2005 [7]	Analytical modeling
Duarte-Melo, Enrique J., and Mingyan Liu, "Analysis of energy consumption and lifetime of heterogeneous wireless sensor networks."	2002 [8]	Analytical modeling
Shah, Rahul C., Sumit Roy, Sushant Jain, and Waylon Brunette, "Data mules: Modeling and analysis of a three-tier architecture for sparse sensor networks."	2003 [9]	Analytical modeling
Kumar, Santosh, Anish Arora, and Ten-Hwang Lai, "On the lifetime analysis of always-on wireless sensor network applications."	2005 [12]	Experiment
da Cunha, Adriano B., and Digenes C. da Silva. "An approach for the reduction of power consumption in sensor nodes of wireless sensor networks: Case analysis of Mica2."	2006 [17]	Experiment
Polastre, Joseph, Robert Szewczyk, Alan Mainwaring, David Culler, and John Anderson. "Analysis of wireless sensor networks for habitat monitoring."	2004 [18]	Experiment
Nguyen, Hoang Anh, Anna Förster, Daniele Puccinelli, and Silvia Giordano. "Sensor node lifetime: An experimental study."	2011 [19]	Experiment
Jung, Deokwoo, Thiago Teixeira, and Andreas Savvides. "Sensor node lifetime analysis: Models and tools."	2009 [10]	Simulation
Di Pietro, Roberto, Luigi V. Mancini, Claudio Soriente, Angelo Spognardi, and Gene Tsudik. "Catch me (if you can): Data survival in unattended sensor networks."	2008 [11]	Simulation
Dron, Wilfried, Simon Duquennoy, Thiemo Voigt, Khalil Hachicha, and Patrick Garda. "An emulation-based method for lifetime estimation of wireless sensor networks."	2014 [20]	Simulation
Ma, Zhanshan, and Axel W. Krings. "Insect population inspired wireless sensor networks: A unified architecture with survival analysis, evolutionary game theory, and hybrid fault models."	2008 [21]	Game Theory & Survival Analysis

formulation helps to quantify the optimal number of clusters and shows how to allocate energy between different layers. In the study of Shah *et al.* [9], simple analytical model is used to analyze the performance of the system. Proposed approach investigates the benefits of three-tier architecture for collecting sensor data. According to given results in the study, three-tier architecture approach can lead to substantial power savings at the sensors.

Another study which analyzes network lifetime by experiment is the study of Kumar *et al.* [12]. The proposed approach is proved by deploying ExScal (a large-scale WSN for intrusion detection) to identify major components in the network lifetime analysis. Results of experiments show how to analyze the effects of using various non-sleep-wakeup power management schemes such as hierarchical sensing, low-power listening, and network data aggregation on the network lifetime. The case study of Cunha *et al.* [17] analyzes wireless sensor node of Mica2 and proposes an approach to reduce power consumption which in turn increases lifetime. Proposed approach is verified with experiment. Another experimental study by Polastre *et al.* [18] analyzes system performance using environmental and node health data from experiment. The study of Nguyen *et al.* [19] also experimentally analyzes lifetime of TelosB sensors using different commercial batteries.

Simulation is another method to analyze WSNs. Jung *et al.* [10] analyze two modes of operation of sensor nodes using models, and the study presents a MATLAB Wireless Sensor Node Platform Lifetime Prediction and Simulation Package (MAT-SNL). Dron *et al.* [20] use Contiki Cooja simulator to analyze and model complex battery characteristics and node lifetimes in WSNs. In the study of Zhanshan *et al.* [21], authors envision a WSN as an entity analogous to a biological population with individual nodes mapping to individual organisms and the network architecture mapping to the biological population. The interactions between individual WSN sensors are captured with evolutionary game theory models. On the node level, survival analysis is introduced to model lifetime, reliability, and survival probability of WSN nodes.

10.3 Data set: Mica2Dot sensors and statistical analysis

The data set which is used in this study is Mica2Dot sensor data from Intel Berkeley Research Lab [22]. The raw data include 2.3 million records of Mica2Dot sensor. Variable names are date, time, epoch (sequence number), mote id, temperature, humidity, light, and voltage. The deployment of sensor nodes are given in Figure 10.1.

To understand and analyze the data, some models are used and results are discussed in details. First, descriptive analysis is conducted. Data set is described and basic statistics such as min, max, mean, and median values for each individual variable are explained. Relationship between lifetime of sensor motes and external factors (e.g. temperature, humidity, and light) is depicted with the help of tables and figures. Considering results of descriptive analysis, regression model is estimated. Results of regression models are discussed and explained in details. After interpreting results of regression model, ordered logit and simple logit models are estimated. Parallel regression assumption of ordered logit models is checked to understand whether it holds or not. Related interpretations are explained and results are discussed.

10.3.1 Descriptive analysis

The data set is a wireless sensor data set with dimension of 1,710,885 observations and 12 variables. Summary of data is given in Table 10.2. "moteid" is the id of each individual mote in the data set. There are 54 motes in total. "temp" variable is the temperature value sent by each sensor mote. Temperature value is in centigrade degrees. "humidity" variable is the air humidity value in percentage between 0 and 100. "light"

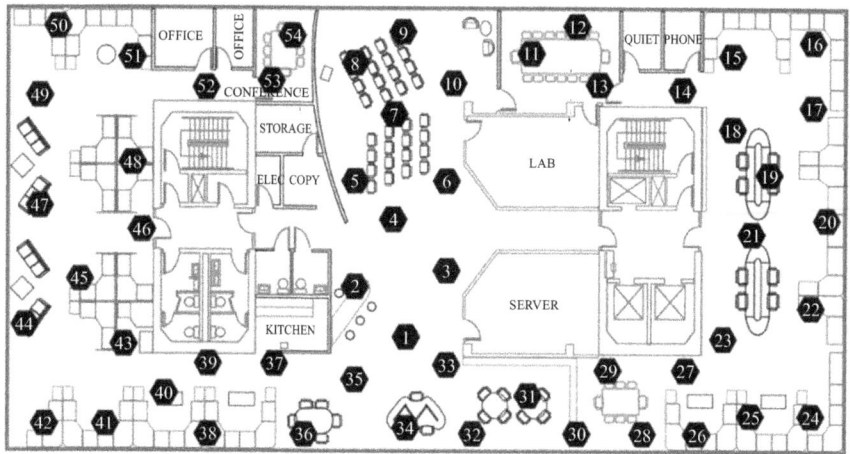

Figure 10.1   Deployment of sensors

Table 10.2   Descriptive statistics of the data

Variable	Min	1st Qu.	Median	Mean	3rd Qu.	Max
moteid	1.00	17.00	30.00	28.76	42.00	54.00
temp	−38.40	20.01	21.79	22.08	23.56	385.57
humidity	−8,983.13	35.20	40.16	39.12	43.52	137.51
light	0.0	43.24	158.24	412.31	555.72	1,847.36
voltage	0.009	2.485	2.593	2.568	2.651	3.159
spell	1	8,352	16,810	18,119	26,480	54,399
dur	0.00	29.00	33.00	54.66	59.00	110,730.00
status	0.00e+00	0.00e+00	0.00e+00	3.04e+05	0.00e+00	1.00e+00
ivoltage	0.000	0.508	0.566	0.591	0.674	3.150
durm	0.000	0.483	0.550	0.911	0.983	1,845.50
durh	0.000	0.008	0.009	0.015	0.016	30.758

is reported in lux. According to data set explanations [22], 1 lux corresponds to moon light, 400 lux to bright office, and 100,000 lux to full sunlight. "voltage" variable is the supply voltage value in volts. It changes between 0 and 3.159. "spell" variable is the number of different spell lengths between each packet transmission. For example, first spell is the time between first and second transmission. "dur" variable is the spell lengths in number of seconds. "status" of the sensor motes shows whether sensor is alive or not. It can be 0 or 1 where 0 is alive and 1 is not alive. "ivoltage" variable is the voltage value of the sensor mote in increasing order. The unit of variable is volts. "durm" is the spell lengths in minutes and "durh" is the spell lengths in hours.

Table 10.2 gives summary of numerical data in data set. By analyzing summary table, it can be seen that there are extreme values for each variable in the data set, which are the sign of malfunctioning of the sensor node since Mica2Dot sensors

Table 10.3   Duration lengths as seconds in groups

0–30	31–60	61–90	91–120	121–150	151–180	181–210	211–240	241–270	270–
681,774	673,222	197,410	75,410	35,044	17,902	10,080	6,043	3,767	10,233

Table 10.4   Increasing voltage in groups

0.00–0.40	0.41–0.67	0.68–
16,949	1,264,339	429,597

Table 10.5   Temperature in groups

0.00–15.00	15.01–24.00	24.01–
1,917	1,347,887	361,081

Table 10.6   Humidity in groups

0–33	34–55	56–
286,134	1,422,623	2,128

work between 2.7 and 3.3 V [23] and it is observed that unexpected measurements are because of low-voltage supply (i.e. less than 2.7 V). To specify the spell ends, the time that sensor node starts to malfunction can be accepted as failure event. In addition, there are missing (NA) values exist. NA value for this data set shows that sensor node stops sending data to the sink, which means that the connection between the sink and sensor nodes is failed and it can be assumed that the first NA value we have is the instance that the sensor node failed or failure event of sensor node.

In Table 10.3, duration variable is split into groups. Number of observations in each group is shown. For example, there are 681,774 observations in the first group in which duration variable is between 0 and 30 seconds. A number of observations are given for other groups as well.

In Table 10.4, increasing voltage variable is split into three groups. Number of observations are given. Tables 10.5, 10.6, and 10.7 show the number of observations in each group for temperature, humidity, and light variables, respectively.

In Figures 10.2, 10.3, 10.4, and 10.5, loess (locally estimated scatterplot smoothing) estimates of temperature, humidity, light, and voltage variables are depicted with respect to duration. Duration is shown in y-axis, and x-axis shows each individual variable which are temperature, humidity, light, and increasing voltage. Temperature variable affects duration variable in negative way since duration decreases as temperature increases. Light variable shows similar pattern with temperature between

Table 10.7   Light in groups

0–50	51–500	501–
463,286	469,699	777,900

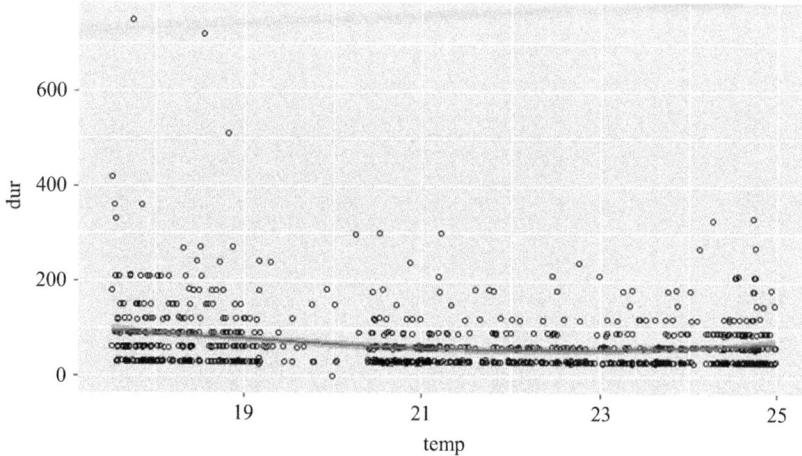

Figure 10.2   Duration by temperature

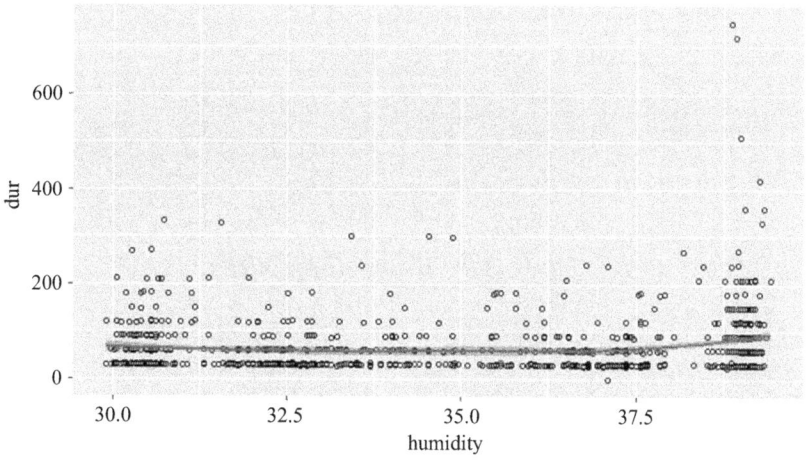

Figure 10.3   Duration by humidity

0 and 200 lux. After 200 lux, light variable affects the duration in positive way in which duration increases as light value increases. On the other hand, humidity and voltage variables show positive relation with duration and lifetime of sensor motes increases as humidity and voltage increase according to results of descriptive analysis.

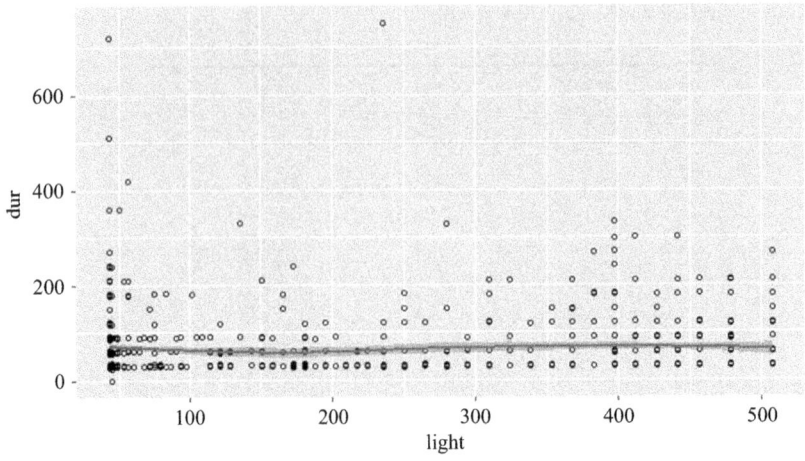

Figure 10.4   Duration by light

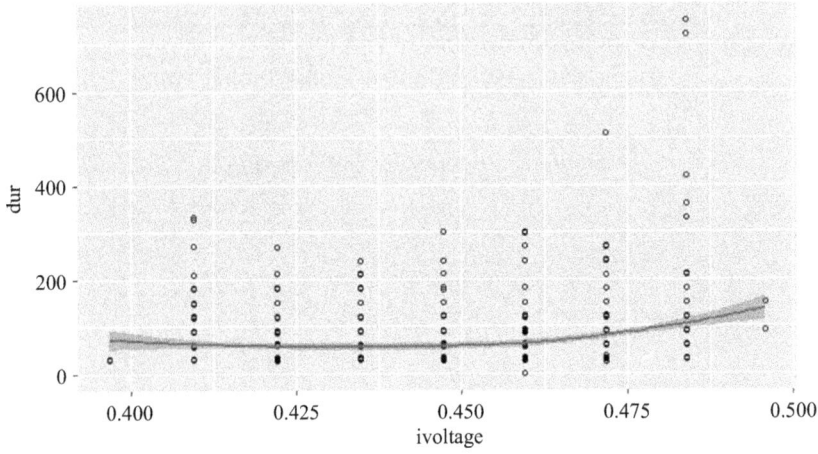

Figure 10.5   Duration by voltage

Regression models are employed in the following sections since descriptive analysis is not sufficient on its own to draw general conclusions. First, linear regression model is estimated. Variables used in the model are explained in details and results of the regression are discussed. As a second model, ordered logit regression is employed. Seven different regression models are compared and discussed in details with necessary figures and tables.

10.3.2   Linear regression model (ordinary least squares estimates)

In this section, employed regression models are discussed. First of all, linear regression model is employed for each independent variable one by one. Some abbreviations

Table 10.8 Temperature

	Dependent variable
	dur
temp	0.238
	(0.150)
Constant	49.408***
	(3.339)
Observations	1,710,885
R^2	0.00000
Adjusted R^2	−0.00000
Residual std. error	619.964 (df = 1,710,883)
F statistic	2.529 (df = 1; 1,710,883)

Note: $^*p < 0.1$; $^{**}p < 0.05$; $^{***}p < 0.01$.

Table 10.9 Humidity

	Dependent variable
	dur
humidity	− 0.254 ***
	(0.051)
Constant	64.583 ***
	(2.044)
Observations	1,710,885
R^2	0.00001
Adjusted R^2	0.00001
Residual std. error	619.960 (df = 1,710,883)
F statistic	24.878***(df = 1; 1,710,883)

Note: $^*p < 0.1$; $^{**}p < 0.05$; $^{***}p < 0.01$.

are used in the following tables. "dur" is the dependent variable in the regression which shows the lifetime (i.e. duration that sensor is alive) of sensor node, and "ivoltage" is the increasing voltage which is used as an independent variable. "R^2" is the statistical measure which shows how close the data are to the regression model. "df" notation means the degrees of freedom. Total degrees of freedom is $n − 1$ which means one less than number of observations and "p" values, on the other hand, shows the significance levels and each significant estimated parameter is shown using stars (*). One star (*) shows significance level of 10%, two stars (**) 5%, and three stars (***) 1%.

According to results, temperature variable does not have significant effect on lifetime. On the other hand, humidity and light have significant effect on lifetime of a sensor. Similarly increasing voltage has significant effect on lifetime as shown in Tables 10.8, 10.9, 10.10, and 10.11.

Table 10.10 Light

	Dependent variable
	dur
light	0.012***
	(0.001)
Constant	49.844***
	(0.595)
Observations	1,710,885
R^2	0.0001
Adjusted R^2	0.0001
Residual std. error	619.932 (df = 1,710,883)
F statistic	179.282*** (df = 1; 1,710,883)

Note: $^*p < 0.1$; $^{**}p < 0.05$; $^{***}p < 0.01$.

Table 10.11 Increasing voltage

	Dependent variable
	dur
ivoltage	38.599***
	(4.725)
Constant	31.873***
	(2.830)
Observations	1,710,885
R^2	0.00004
Adjusted R^2	0.00004
Residual std. error	619.952 (df = 1,710,883)
F statistic	66.745*** (df = 1; 1,710,883)

Note: $^*p < 0.1$; $^{**}p < 0.05$; $^{***}p < 0.01$.

After regression results of each independent variable, using temperature, humidity, light, and increasing voltage variables as independent variable another regression model is estimated. Dependent variable is duration again. The results are depicted as shown in Table 10.12. According to results, coefficients of temperature and humidity variables are significant and affect duration variable in negative way. On the other hand, coefficients of light and increasing voltage variables are also significant but affect the lifetime of sensor motes in positive way.

10.3.3 Ordered logit models

In this section, ordered logit model is employed. In ordered logit regression model, grouped duration variable is used as dependent variable. In the first model, all groups are used together. Dependent variable is duration groups (durG), and independent

Table 10.12   *Temperature + humidity + light + increasing voltage*

	Dependent variable
	dur
temp	−1.056***
	(0.185)
humidity	−0.399***
	(0.058)
light	0.014***
	(0.001)
ivoltage	49.613***
	(4.906)
Constant	58.533***
	(6.027)
Observations	1,710,885
R^2	0.0002
Adjusted R^2	0.0002
Residual std. error	619.905 (df = 1,710,880)
F Statistic	82.335*** (df = 4; 1,710,880)

Note: $^*p < 0.1$; $^{**}p < 0.05$; $^{***}p < 0.01$.

variables are temperature, humidity, light, and increasing voltage. According to results of the model, the coefficient of temperature variable is significant and the sign of the coefficient is negative. On the other hand, coefficients of humidity and light variables are also significant but the sign of the coefficients is positive. In this model, the coefficient of increasing voltage is not significant as shown in Table 10.13.

In Table 10.14, threshold coefficients for ordered logit model are depicted. Estimates, standard errors, and z values of each group threshold are reported. In Table 10.15, the means of predicted probabilities of ordered logit model are shown as well.

Ordered model estimates assume parallel impacts of explanatory variables on duration groups. Simple logit model estimates give possibility to test this assumption. Seven simple logit models are estimated to test the assumption. In Table 10.16, ordered logit regression results of the first model is shown. According to the results, the coefficient of temperature variable is significant and affects the duration variable positively. Coefficients of humidity and light variables are also significant and positively affect the duration variable. On the other hand, the coefficient of increasing voltage variable is also significant but it affects the duration variable negatively.

In Table 10.17, ordered logit regression results of the second model are depicted. According to the results, the coefficient of temperature variable is significant and affects the duration variable positively. Coefficients of humidity and light variables are also significant and positively affect the duration variable. On the other hand,

Table 10.13　Ordered logit model

	Dependent variable
	durG
scale (temp)	−0.009***
	(0.002)
scale (humidity)	0.074***
	(0.003)
scale (light)	0.018***
	(0.002)
scale (ivoltage)	−0.001
	(0.002)
Observations	1,710,885
Log-likelihood	−2,295,436.000

Note: $^*p < 0.1$; $^{**}p < 0.05$; $^{***}p < 0.01$.

Table 10.14　Threshold coefficients of ordered logit model

	Estimate	**Std. error**	**z Value**
0–30—31–60	−0.412168	0.001562	−263.9
31–60—61–90	1.337506	0.001884	709.9
61–90—91–120	2.282889	0.002638	865.5
91–120—121–150	2.976448	0.003557	836.7
121–150—151–180	3.545796	0.004629	766.0
151–180—181–210	4.022997	0.005813	692.0
181–210—211–240	4.436423	0.007105	624.4
211–240—241–270	4.798839	0.008487	565.5
241–270—270–	5.114514	0.009915	515.8

Table 10.15　The means of predicted probabilities of ordered logit model

0–30	31–60	61–90	91–120	121–150	151–180
0.398780656	0.393498447	0.115558469	0.044189460	0.020555703	0.010508681
181–210	211–240	241–270	270–		
0.005925281	0.003551944	0.002213760	0.006036991		

coefficient of increasing voltage variable is also significant but it affects the duration variable negatively.

In Table 10.18, ordered logit regression results of model three are reported. According to the results, the coefficient of temperature variable is significant and

Table 10.16   Model 1

	Dependent variable
	I(as.numeric(durG) ≥ 1)
scale (temp)	0.017***
	(0.002)
scale (humidity)	0.088***
	(0.004)
scale (light)	0.006***
	(0.002)
scale (ivoltage)	−0.003*
	(0.002)
Constant	0.412***
	(0.002)
Observations	1,710,885
Log-likelihood	−2,295,436.000
Akaike Inf. Crit.	2,299,869.000

Note: *$p < 0.1$; **$p < 0.05$; ***$p < 0.01$.

Table 10.17   Model 2

	Dependent variable
	I(as.numeric(durG) ≥ 2)
scale (temp)	0.017***
	(0.002)
scale (humidity)	0.088***
	(0.004)
scale (light)	0.006***
	(0.002)
scale (ivoltage)	−0.003*
	(0.002)
Constant	0.412***
	(0.002)
Observations	1,710,885
Log-likelihood	−2,295,436.000
Akaike Inf. Crit.	2,299,869.000

Note: *$p < 0.1$; **$p < 0.05$; ***$p < 0.01$.

affects the duration variable negatively. Coefficients of humidity and light variables are also significant and positively affect the duration variable. On the other hand, coefficient of increasing voltage variable is also significant but it affects the duration variable negatively.

In Table 10.19, ordered logit regression results of model four are reported. According to the results, the coefficient of temperature variable is significant and

Table 10.18   Model 3

	Dependent variable
	I(as.numeric(durG) ≥ 3)
scale (temp)	−0.049***
	(0.003)
scale (humidity)	0.057***
	(0.005)
scale (light)	0.035***
	(0.002)
scale (ivoltage)	−0.005**
	(0.002)
Constant	−1.339***
	(0.002)
Observations	1,710,885
Log-likelihood	−874,071.600
Akaike Inf. Crit.	1,748,153.000

Note: $^{*}p < 0.1$; $^{**}p < 0.05$; $^{***}p < 0.01$.

Table 10.19   Model 4

	Dependent variable
	I(as.numeric(durG) ≥ 4)
scale (temp)	−0.090***
	(0.004)
scale (humidity)	0.030***
	(0.006)
scale (light)	0.057***
	(0.003)
scale (ivoltage)	0.013***
	(0.003)
Constant	−2.286***
	(0.003)
Observations	1,710,885
Log-likelihood	−874,071.600
Akaike Inf. Crit.	1,748,153.000

Note: $^{*}p < 0.1$; $^{**}p < 0.05$; $^{***}p < 0.01$.

affects the duration variable negatively. On the other hand, coefficients of humidity, light, and increasing voltage variables are also significant and positively affect the duration variable.

In Table 10.20, ordered logit regression results of fifth model are shown. According to the results, coefficient of temperature variable is significant and affects the

Table 10.20   Model 5

	Dependent variable
	I(as.numeric(durG) ≥ 5)
scale (temp)	−0.123***
	(0.005)
scale (humidity)	0.006
	(0.007)
scale (light)	0.080***
	(0.004)
scale (ivoltage)	0.044***
	(0.004)
Constant	−2.983***
	(0.004)
Observations	1,710,885
Log-likelihood	−331,675.500
Akaike Inf. Crit.	663,361.000

Note: $^{*}p < 0.1$; $^{**}p < 0.05$; $^{***}p < 0.01$.

Table 10.21   Model 6

	Dependent variable
	I(as.numeric(durG) ≥ 6)
scale (temp)	−0.144***
	(0.006)
scale (humidity)	0.001
	(0.003)
scale (light)	0.106***
	(0.005)
scale (ivoltage)	0.074***
	(0.005)
Constant	−3.557***
	(0.005)
Observations	1,710,885
Log-likelihood	−218,338.700
Akaike Inf. Crit.	436,687.300

Note: $^{*}p < 0.1$; $^{**}p < 0.05$; $^{***}p < 0.01$.

duration variable negatively. On the other hand, coefficients of light and increasing voltage variables are also significant and positively affect the duration variable. In this model, coefficient of humidity variable loses its significance.

In Table 10.21, ordered logit regression results of model six are reported. According to the results, the coefficient of temperature variable is significant and it affects

Table 10.22　Model 7

	Dependent variable
	I(as.numeric(durG) ≥ 7)
scale (temp)	−0.158***
	(0.007)
scale (humidity)	0.001
	(0.002)
scale (light)	0.135***
	(0.006)
scale (ivoltage)	0.117***
	(0.006)
Constant	−4.042***
	(0.006)
Observations	1,710,885
Log-likelihood	−150,924.800
Akaike Inf. Crit.	301,859.600

Note: $^*p < 0.1$; $^{**}p < 0.05$; $^{***}p < 0.01$.

the duration variable negatively. On the other hand, coefficients of light and increasing voltage variables are also significant and positively affect the duration variable. However, the coefficient of humidity variable is not significant.

In Table 10.21, ordered logit regression results of model seven (last model) are shown. According to the results, the coefficient of temperature variable is significant and affects the duration variable negatively. On the other hand, coefficients of light and increasing voltage variables are also significant and positively affect the duration variable. However, the coefficient of humidity variable is not significant.

In Table 10.23, ordered logit models are summarized. Seven different models are shown in the table. For each independent variable, sign of significance is depicted. "+" (positive) sign shows that the coefficient of variable is significant and affects the dependent variable in positive way. "–" (negative) sign means that the coefficient of variable is significant and affects the dependent variable in negative way. "0" (zero) means that coefficient of the variable is not significant. By comparing the sign of each variable, parallel regression assumption of ordered logit models is checked. If it holds then simple logit estimates of each separate logit model have the similar estimated parameter values; that is, for each group parameter, estimates have similar sign, size, and significance. Considering summary table, it is shown that parallel regression assumption of ordered logit models does not hold since sign, size, or significance of variables change for different ordered logit models. For example, the sign of temperature variable is positive for Models 1 and 2; however, it changes to negative for Model 3 and remaining models. Similarly, the sign of humidity variable is positive until Model 4, but it loses its significance after Model 5. Similar sign change is observed for increasing voltage variable after Model 3.

Table 10.23   Ordered logit models

	M1	M2	M3	M4	M5	M6	M7
temp	+	+	−	−	−	−	−
humidity	+	+	+	+	0	0	0
light	+	+	+	+	+	+	+
ivoltage	−	−	−	+	+	+	+

Table 10.24   Logit model: odds ratio

	Estimate	OR
(Intercept)	0.4119	1.5097
scale (temp)	0.0173	1.0175
scale (humidity)	0.0881	1.0921
scale (light)	0.0065	1.0065
scale (ivoltage)	−0.0029	0.9971

Table 10.25   Marginal effects

	dF/dx	Std. err.	z	$P > |z|$
scale (temp)	0.00414654	0.00058169	7.1284	1.015e-12 ***
scale (humidity)	0.02111267	0.00089949	23.4718	2.2e-16 ***
scale (light)	0.00155306	0.00041759	3.7191	0.0001999 ***
scale (ivoltage)	−0.00070484	0.00041786	−1.6868	0.0916467·

The Estimate column in Table 10.24 shows the coefficients in log-odds form. When "Temperature" increases by one unit, the expected change in the log of odds is 0.0173. What we get from this column is whether the effect of the predictors is positive or negative. Based on the output, when "temp" increases by one unit, the odds of $durG \geqslant 2$ increase by 51% $(1.5097 - 1) * 100$. Or, the odds of $durG \geqslant 2$ are 1.51 times higher when "temp" increases by one unit (keeping all other predictors constant).

In Table 10.25, marginal effects are depicted. Marginal effects show the change in probability when the predictor or independent variable increases by one unit. For continuous variables, this represents the instantaneous change given that the "unit" may be very small. For binary variables, the change is from 0 to 1, so one "unit" as it is usually thought.

Since the main aim of the study is explaining variables which have effects on lifetime of low-power sensors, descriptive analysis of raw data is conducted to understand data well. After descriptive statistics, data are analyzed using linear regression model. According to results of linear regression model, it is shown that linear regression does not explain low-power sensor data well. After linear regression model, ordered logit regression is employed and results are discussed. As a next step, survival analysis will be employed and it is expected that it will fit the low-power sensor data better where the main variables are lifetime and the factors affecting it.

10.4   Conclusion and future works

The concept of IoT and smart environment applications are becoming more and more popular. With smartness of environment and wide use of electronic devices and sensors, lifetime and energy consumption analysis of WSNs became a must. Wide application area of WSNs such as disaster surveillance and health care raises the importance of lifetime analysis. In this study, statistical analysis is conducted for lifetime of wireless sensors using the data from *"Intel Berkeley Research Lab."* As a first step, literature review is conducted to show which methods are used to analyze sensor lifetime. The corresponding categorization is done and presented in previous sections. Generally, there are three main methods in the literature which are analytical modeling, experiment, and simulation. Statistical analysis is rare in the area of WSNs in terms of lifetime. After literature review, data set is described and explained in details by descriptive analysis. Results of descriptive analysis are proposed and discussed. Moreover, linear regression and ordered logit regression analysis are conducted. The results of analysis are also discussed. In conclusion, the importance of statistical analysis to understand variables that affect lifetime is presented. In addition to well-known methods, statistical analysis not only estimates the lifetime but also determines the important variables that affect lifetime of wireless sensor nodes. As a future work or second part of study, data will be arranged to be more compatible with survival methods and the analysis will be conducted to see relations or effects of each individual parameter on lifetime of a low-power sensors.

References

[1]   Arasteh H, Hosseinnezhad V, Loia V, *et al.* IoT-based smart cities: A survey. In: 2016 IEEE 16th International Conference on Environment and Electrical Engineering (EEEIC). IEEE; 2016. pp. 1–6.

[2]   Akyildiz IF, Melodia T, and Chowdhury KR. A survey on wireless multimedia sensor networks. Computer Networks. 2007;51(4):921–960.

[3]   Akyildiz IF, Su W, Sankarasubramaniam Y, *et al.* Wireless sensor networks: A survey. Computer Networks. 2002;38(4):393–422.

[4]   Bekaroo G, Bokhoree C, and Pattinson C. Impacts of ICT on the natural ecosystem: A grassroot analysis for promoting socio-environmental sustainability. Renewable and Sustainable Energy Reviews. 2016;57:1580–1595.

[5]   Polastre J, Szewczyk R, and Culler D. Telos: Enabling ultra-low power wireless research. In: Proceedings of the 4th International Symposium on Information Processing in Sensor Networks. IEEE Press; 2005. p. 48.

[6]   Chelloug SA. Impact of the temperature and humidity variations on link quality of xm1000 mote sensors. arXiv preprint arXiv:150101073. 2015.

[7]   Chen Y, and Zhao Q. On the lifetime of wireless sensor networks. IEEE Communications Letters. 2005;9(11):976–978.

[8]   Duarte-Melo EJ, and Liu M. Analysis of energy consumption and lifetime of heterogeneous wireless sensor networks. In: Global Telecommunications Conference, 2002. GLOBECOM'02. IEEE. Vol. 1. IEEE; 2002. pp. 21–25.

[9]   Shah RC, Roy S, Jain S, *et al.* Data mules: Modeling and analysis of a three-tier architecture for sparse sensor networks. Ad Hoc Networks. 2003;1(2-3): 215–233.

[10]   Jung D, Teixeira T, and Savvides A. Sensor node lifetime analysis: Models and tools. ACM Transactions on Sensor Networks (TOSN). 2009;5(1):3.

[11]   Di Pietro R, Mancini LV, Soriente C, *et al.* Catch me (if you can): Data survival in unattended sensor networks. In: 2008 Sixth Annual IEEE International Conference on Pervasive Computing and Communications (PerCom). IEEE; 2008. pp. 185–194.

[12]   Kumar S, Arora A, and Lai TH. On the lifetime analysis of always-on wireless sensor network applications. In: IEEE International Conference on Mobile Ad Hoc and Sensor Systems Conference, 2005. IEEE; 2005. pp. 183–188.

[13]   Kizilkaya B, Caglar M, Al-Turjman F, *et al.* Binary search tree based hierarchical placement algorithm for IoT-based smart parking applications. Internet of Things. 2019;5:71–83.

[14]   Al-Turjman F. Guest Editorial: Smart cities and smart sensory platforms. Inst Engineering Technology-IET Michael Faraday House Six Hills Way; 2018. pp. 247–248.

[15]   Al-Turjman F, Hasan MZ, and Al-Rizzo H. Task scheduling in cloud-based survivability applications using swarm optimization in IoT. Transactions on Emerging Telecommunications Technologies. 2018:10.1002/ett.3539.

[16]   Pino T, Choudhury S, and Al-Turjman F. Dominating set algorithms for wireless sensor networks survivability. IEEE Access. 2018;6:17527–17532.

[17]   da Cunha AB, and da Silva DC. An approach for the reduction of power consumption in sensor nodes of wireless sensor networks: Case analysis of Mica2. In: International Workshop on Embedded Computer Systems. Springer; 2006. pp. 132–141.

[18]   Polastre J, Szewczyk R, Mainwaring A, *et al.* Analysis of wireless sensor networks for habitat monitoring. In: Wireless Sensor Networks. Boston, MA: Springer; 2004. pp. 399–423.

[19]   Nguyen HA, Förster A, Puccinelli D, *et al.* Sensor node lifetime: An experimental study. In: 2011 IEEE International Conference on Pervasive Computing and Communications Workshops (PERCOM Workshops). IEEE; 2011. pp. 202–207.

[20]   Dron W, Duquennoy S, Voigt T, *et al.* An emulation-based method for life-time estimation of wireless sensor networks. In: 2014 IEEE International Conference on Distributed Computing in Sensor Systems. IEEE; 2014. pp. 241–248.

[21]   Ma Z, and Krings AW. Insect population inspired wireless sensor networks: A unified architecture with survival analysis, evolutionary game theory, and hybrid fault models. In: 2008 International Conference on BioMedical Engineering and Informatics. Vol. 2. IEEE; 2008. pp. 636–643.

[22]   Madden S. Intel lab data. Web page, Intel. 2004.

[23]   Datasheet M. Crossbow Technology Inc. San Jose, California. 2006; p. 50.

Appendix A
Variables by duration groups

In Table A.1, number of observations in each voltage group is shown by each duration group. Tables A.2–A.4 show the number of observations in each group for temperature, humidity, and light variables, respectively, for each duration group.

Table A.1 Voltage by duration

	0–30	31–60	61–90	91–120	121–150	151–180	181–210	211–240	241–270	270–
0.00–0.40	7,481	6,566	1,734	614	278	115	57	35	24	45
0.41–0.67	503,753	499,088	146,239	55,937	25,780	13,123	7,237	4,269	2,638	6,275
0.68–	170,540	167,568	49,437	18,859	8,986	4,664	2,786	1,739	1,105	3,913

Table A.2 Temperature by duration

	0–30	31–60	61–90	91–120	121–150	151–180	181–210	211–240	241–270	270–
00.00–15.00	666	586	279	148	75	48	37	27	14	37
15.01–24.00	533,613	530,252	156,800	60,226	28,213	14,455	8,094	4,898	2,988	8,348
24.01–	147,495	142,384	40,331	15,036	656	3,399	1,949	1,118	765	1,848

Table A.3 Humidity by duration

	0–30	31–60	61–90	91–120	121–150	151–180	181–210	211–240	241–270	270–
0–33	118,626	110,177	31,708	12,100	5,594	2,828	1,627	1,023	649	1,802
34–55	562,284	562,273	165,442	63,209	29,395	15,038	8,436	5,013	3,113	8,420
56–	864	772	260	101	55	36	17	7	5	118

Table A.4 Light by duration

	0–30	31–60	61–90	91–120	121–150	151–180	181–210	211–240	241–270	270–
0–50	179,169	184,126	55,285	21,043	9,990	5,140	2,888	1,715	1,026	2,904
51–500	313,867	304,201	88,852	34,193	15,695	8,096	4,518	2,633	1,661	4,184
501–	188,738	184,895	53,273	20,174	9,359	4,666	2,674	1,695	1,080	3,145

Chapter 11
Conclusions and recommendations
Fadi Al-Turjman

Smart cities have emerged as one of the most promising wireless sensor networks (WSNs) applications in the Internet of Things (IoT) era because of their agile nature and significant impact in almost everyday activity. Toward more efficient smart cities implementations, in this work, we proposed and evaluated the use of sensors and WSNs in large-scale IoT applications. We focused on key design aspects in the sensor node and WSNs performance, deployment and data readings trends. In the following is a summary of our conclusions and recommendations for the future relevant work.

We started the work in this book with a comprehensive overview about the smart cities in the IoT era, while focusing on security issues in **Chapter 1**. In this chapter, we have overviewed smart city applications, the IoT and CPSs as enabling technologies of smart cities, and the relation between the two. The technological constituents of the smart city infrastructure need to be integrated and they interact with each other for provisioning services to citizens. In addition, each enabling technology has its own unique characteristics. The interplay among smart city constituents and their unique characteristics pose unique security and privacy challenges. In this chapter, we have presented a high-level overview of the security and privacy challenges of smart cities and the unique characteristics of smart cities that make these challenges significant hurdles for the success of smart cities. Furthermore, we discussed some AC approaches from the literature which can be potential solutions to alleviate the security issues. The approaches we present include both the ones proposed specifically for smart cities and their enabling technologies, and the ones that address security risk, trust and secure interoperation issues. We observed that there is a wealth of AC models and frameworks built on role-based approach due to its maturity, whereas attribute-based AC domain needs more exploration. In this direction, we expect that secure interoperability of ABAC policies is a potential research direction.

In **Chapter 2**, we conclude that IoT is not only providing the facilities for smart cities in a more efficient manner, but it also makes the quality of the citizen's life better. The implementation of IoT in all the urban areas will take time and effort but will be a profitable one. The governments are coming forward all over the globe to make it market-friendly for the startups that are working on the IoT smart city solutions. Many smart city expos are being conducted to explore as well as to make aware of IoT usage all over the world. It will enhance the quality, performance and interactivity of the services provided by the city thus reducing cost by utilization of resources in an optimized manner.

In **Chapter 3**, a novel IoT-based smart water layer-based model is discussed which focuses on smart water management through seven layers. The very principle of every advancement is helping society provided, without disturbing the nature's law. In the proposed structure, that principle is carefully considered, and the same time, the efficient utilization of natural resources is also taken care of. Each of the layers is highlighting the various factors which are influencing the smart water decision. The water is a basic need of every living being on earth and unfortunately, till date, the science is incapable to bring out an alternative method to produce water. Hence, this non-renewable energy needs to be invested very carefully so that the future generation will not suffer from water scarcity. At the same time, the other side of science is enough matured to provide smart solutions to manage the water resource and to avoid water wastage. Personally, we feel the coverage area of the water distribution network is so large, and it is very difficult to manage efficiently. Hence, emerging technologies like IoT, ICT and WSN could be the solution. This chapter helps novice researcher in the water domain to understand the insights of water-related problems and challenges.

In **Chapter 4**, we discuss three algorithms, their merits and demerits. First, the proposed scheme scans the generated log file and provides summary of all the IoT motes in separate files. This technique is useful for very large files and complex operation although it requires extended hard disk space for temporary files. Second, the proposed algorithm scans log file to summarize data in memory. This algorithm requires additional space for temporary files and scans source files many times, and consequently, it requires more time to complete the evaluation. Third algorithm scans log file exactly once, does not require any additional space for temporary files and computes summaries in memory. It makes processing really fast and can work without temporary files generated. All three algorithms are helpful in different IoT deployment scenarios; therefore, researcher can choose according to their preference of memory requirements, file sizes and time constraints.

Chapter 5 analyzes the security aspects of the mesh IoT. After introducing the IoT mesh theme, a general analysis of the safety of the IoT was presented. Next, security aspects in the IEEE 802.15.4 standard were analyzed. Significant attention was also paid to the technical guidelines that enable secure transmission of information between selected IoT mesh points. The security of the implementation of the mesh IoT network was analyzed in detail. Finally, the security aspects of different systems were compared.

In **Chapter 6**, smart sensors mounted in the drainage and sewer pipes have been used to read the flow velocity and alert once the flow reaches a velocity in which sediment deposition is occurred. In order to determine the sediment deposition velocity, this study models sediment transport in drainage systems by means of evolutionary decision tree (EDT) technique. EDT results are compared with conventional decision tree (DT) and evolutionary genetic programming (GP) techniques. A large number of experimental data covering wide ranges of sediment and pipe size were used for the modeling. Evaluation of the developed models in terms of verity of statistical indices showed the outperformance of the proposed EDT model. The EDT, DT and GP models were found superior to their traditional corresponding regression models existing in the literature. Results are helpful for determination of the flow characteristics at sediment deposition condition in drainage systems maintained using IoT technology in smart cities.

In **Chapter 7**, we focused on understanding activities of daily life (ADL) from the unobtrusive binary sensors deployed in a smart home. Even though there are many applications that do ADL recognition with numerous methods, we indicated that understanding the data well is essential to provide higher accuracy and to have better control on the system. To emphasize these, we select an exemplary data set to present the possible problems faced during the preprocessing stage. Lastly, we present a statistical method along with an additional feature extracted from the time information of the data, i.e., period of the day, that can be used either for classification or for enhancing the existing models for ADL recognition along with conditional random field and dynamic Bayesian network.

In **Chapter 8**, we provide a comprehensive understanding of the existing cybersecurity attacks on IoT medical devices. We provide a categorization of the cybersecurity attacks on medical IoT devices which have been seen in the wild and can cause security issues and challenges in smart city healthcare services. Moreover, we present security mechanisms, derived from the literature for the most common attacks. We also highlight the emerging good practice and approaches that manufacturers can take in order to improve the medical IoT device security throughout its life cycle. Thus, we provide a foundation for organizing research efforts toward the development of proper security mechanisms against cyberattacks targeting IoT medical devices.

In **Chapter 9**, we overview two potential authentication mechanisms, namely Centralized Authentication Control and Distributed Authentication Control, in detail. Later, we discuss another authentication mechanism, known as hybrid slotted-CSMA/CA (HSCT), as a case study that is proposed to overcome the aforementioned methods' issues. The HSCT mechanism allows IoT systems in smart cities to register thousands of low-power IoT devices (sensors and actuators). This chapter also comes up with the analyses of the access period in a single HSCT time slot.

In **Chapter 10**, a real data set from "Intel Berkeley Research Lab" has been used in lifetime analysis of the used wireless sensor nodes in smart cities. As a first step, a literature review is conducted to show which methods are used to analyze the sensor lifetime. Generally, there are three main methods in the literature which can be categorized into analytical, experimental and simulation-based methods. Statistical analysis is rare in the area of WSNs in terms of lifetime. Therefore, the targeted data set has been described and explained in detail using descriptive analysis. Results of the descriptive analysis are proposed and discussed. Moreover, linear regression and ordered logit regression analysis have been conducted. The results of this analysis are also discussed. In conclusion, the importance of statistical analysis, which is necessary to understand the main variables that affect the sensor lifetime, has been presented. As a future work or second part of study, data will be arranged to be more compatible with survival analysis and the analysis will be conducted to see relations or effects of each individual parameter on lifetime of a low-power sensors.

Index